The Windscale Inquiry

Report by
the Hon. Mr. Justice Parker

Presented to the Secretary of State
for the Environment on 26 January 1978

London : Her Majesty's Stationery Office

©*Crown copyright 1978*
First published 1978

Vol 2 Annexes 1 and 2 and alphabetical index to list of documents.

Vol 3 Index to the Report will be published separately

ISBN 0 11 751314 8

Cumbria County Council
Copeland Borough Council
Application
by
British Nuclear Fuels Limited

Inspector: The Hon Mr Justice Parker

Assessors: Sir Edward Pochin CBE MD FRCP
Professor Sir Frederick Warner C.Eng FRS

Dates of Inquiry: 14 June — 19 October and
24 October — 4 November 1977

Table of Contents

		Para Nos. Inclusive
1	**Introduction**	1.1–1.7
2	**Essential Background**	
	Nuclear reactors and their fuel	2.2
	Radioactivity and radiation	2.13
	Reprocessing	2.20
	Magnox spent fuel	2.22
	Oxide spent fuel	2.27
	Oxide fuel arisings to the year 2000	2.30
	BNFL's capability without THORP	2.33
	Plutonium	2.34 complete
3	**History of the Application**	3.1–3.3
4	**Scope of the Inquiry**	4.1–4.2
5	**Summary of Contentions and Structure of the Report**	
	Applicant's case	5.1
	Objectors' cases	5.2–5.4
6	**The Nuclear Weapons Proliferation Question**	6.1–6.34
7	**Terrorism and Civil Liberties**	7.1–7.25
8	**The Need for Reprocessing of Oxide Fuel and Relationship to the FBR Question**	8.1–8.56
9	**Financial Aspects**	9.1–9.20
10	**Routine Discharges – Risks**	
	System of protection	10.1
	Risk levels if THORP is built	10.18
	Risk levels – suggested inadequacies of current estimates and limits	10.41
	Dr Alice Stewart	10.42
	Professor Edward Radford	10.48
	Dr Sadao Ichikawa	10.53
	Professor William Potts	10.54
	Dr J K Spearing	10.58
	Professor Ellis	10.60
	Professor Ivan Tolstoy	10.65
	Dr V T Bowen	10.66
	Tests made during the inquiry	10.67
	Manchester Water Supply	10.69
	Isle of Man Potatoes	10.72
	Scallops	10.73
	Air at Ravenglass	10.76

		Para Nos. Inclusive
	Whole body monitoring of fish eaters for Caesium	10.85
	Blistered fish	10.96
	Radioactive furniture	10.99
	Is the system defective?	
	The International Aspect	10.101
	The National Aspect	10.109
	Public participation in the control system	10.118
	Compensation for harm done by radiation	10.123
	Monitoring	10.126
	Research	10.127
	Quality and integrity of the advisory and control authorities	10.130
	Miscellaneous	10.139–10.140
11	**Risks – Accidents**	
	General	11.1
	Risks not involving releases beyond the perimeter	11.9
	Accidents involving releases beyond the site	11.12
	Accident risks – industrial action	11.19
	Adequacy of the system	11.24
	Risks during transport	11.25
	The emergency plan and local liaison committee	11.29–11.35
12	**Size of Plant**	12.1–12.4
13	**Public Hostility**	13.1–13.9
14	**Conventional Planning Issues**	
	General	14.1
	Suitability of the site	14.5
	Effect on amenity	14.11
	Visual impact	14.11
	Noise and nuisance	14.17
	Implications for local employment	14.18
	Training	14.25
	Traffic movements	14.26
	Housing	14.31
	Water supplies, sewerage and sewage treatment	14.34
	Financing of housing and infrastructure improvements	14.36
	Conclusions and recommendations on planning issues	14.38–14.45

		Para Nos. Inclusive
15	**The Inquiry Itself**	
	Time interval between the announcement and the opening of the inquiry	15.2
	Location of the inquiry	15.3
	Programming	15.4
	The identity of my assessors	15.7
	Financial disparity	15.8
	Costs of the inquiry	15.10
	Nature of the inquiry	15.12–15.15
16	**Overall Conclusion and Recommendation**	16.1–16.2

		Para Nos. Inclusive
17	**Summary of Principal Conclusions and Recommendations**	
	Conclusions	17.1
	Recommendations	17.8 complete
18	**Miscellaneous Matters**	
	Inquiry procedure	18.1
	Written representations	18.2
	Acknowledgements	18.3 complete

Annex 1 – List of Appearances ⎫ contained in a
Annex 2 – List of Documents ⎭ separate volume
Annex 3 – Somatic Risk Estimates
Annex 4 – Genetic Risk Estimates
Annex 5 – Mean Genetic Dose to Local Population

To the Rt Hon Peter Shore MP, Secretary of State for the Environment

1 Introduction

1.1 On 1 March 1977 British Nuclear Fuels Limited (BNFL) submitted to the Copeland Borough Council (Copeland) an application (BNFL4) under Section 23 of the Town and Country Planning Act 1971 (the 1971 Act) for outline planning permission for 'a plant for reprocessing irradiated oxide nuclear fuels and support site services' at their Windscale and Calder Works, Sellafield, Cumbria. The proposed plant is described in this report as THORP (*T*hermal *O*xide *R*eprocessing *P*lant). This application was referred by Copeland to the Cumbria County Council (Cumbria) as a county matter and called in under Section 35 of the 1971 Act on 25 March 1977. On 31 March 1977, I was appointed to conduct a local inquiry for the purpose of hearing objections and representations relating to the proposed development. There were appointed as my assessors Sir Edward Pochin, CBE MD FRCP and Professor Sir Frederick Warner, C.Eng, FRS.

1.2 The points which, pursuant to Rule 6(1) of the Town and Country Planning (Inquiries Procedure) Rules 1974 (the Procedure Rules), were stated to be likely to be relevant to the consideration of BNFL's application were:
  i. the implications of the proposed development for the safety of the public and for other aspects of the national interest;
  ii. the implications for the environment of the construction and operation of the proposed development in view of the measures that can be adopted under:
    a. the Radioactive Substances Act 1960 to control the disposal of solid, gaseous and liquid wastes which would result from the proposed development;
    and
    b. the Nuclear Installations Act 1965 to provide for the safety of operations at the reprocessing plant;
  iii. the effect of the proposed development on the amenities of the area;
  iv. the effect of additional traffic movements both by road and rail and which would result from the proposed development;
  v. the implications of the proposed development for local employment;
  vi. the extent of the additional provision that would need to be made for housing and public services as a result of the proposed development.

1.3 I held a preliminary meeting on 17 May 1977 at the Civic Hall, Whitehaven, Cumbria, in order to discuss and settle procedural matters. The Inquiry was opened at the same place on Tuesday, 14 June 1977. It closed on Friday 4 November 1977, the one-hundredth day of the hearings. I visited the site on the day before the Inquiry opened and BNFL's solid waste disposal site at Drigg immediately following its close.

1.4 During the hearings evidence was taken from 146 witnesses and a transcript of all the evidence was provided by a team of shorthand writers. The speed and accuracy with which the daily transcripts were produced was of the greatest assistance and received many well-deserved tributes from the parties. To those tributes I add my own. In addition to the oral evidence a large number of documents – many of them books – were put in. They totalled some 1,500. Lists of witnesses and documents are appended as Annexes 1–2*. Five films were also shown to me on behalf of objectors.

1.5 From time to time during the course of the Inquiry certain tests and research work were carried out at my request. I shall refer to these in detail later in this report. I mention them at the outset for I wish to express at once my appreciation to all those who carried them out or participated in them, particularly to the inhabitants of the village of Ravenglass, whose co-operation enabled the National Radiological Protection Board (NRPB) to carry out air sampling in the village over a period of one month, and to a number of local residents, who arranged and submitted themselves to tests to determine what was their body content of radioactive caesium as a result of their consumption of fish caught in the Irish Sea close to the site.

1.6 At one stage it was, somewhat tentatively, suggested on behalf of the Isle of Man Government (IoM), that there were defects in the application itself or the steps leading up to the Inquiry. This suggestion was however abandoned by their Counsel in his closing submissions. No other party suggested any such defects in the procedural steps required by law although, as will appear later, submissions were made that changes in the procedure would, in certain respects, be desirable.

1.7 At the outset of the Inquiry I posed three questions which appeared to me to be sufficient to cover all issues which had then been indicated. These questions were:
1. Should oxide fuel from United Kingdom reactors be reprocessed in this country at all, whether at

---

* Contained in Volume 2.

Windscale or elsewhere?
2. If yes, should such reprocessing be carried on at Windscale?
3. If yes, should the reprocessing plant be about double the estimated size required to handle United Kingdom oxide fuels and be used, as to the spare capacity, for reprocessing foreign fuels?

These three questions still appear to me to cover all the issues raised at the Inquiry, numerous as they were. The applicants submitted that the answers to all three questions should be in the affirmative. Other parties' submissions ranged from an outright 'No' to all three questions through various permutations including 'Not yet', 'Yes' to questions 1–2, but 'No' to question 3 and 'Yes but subject to preconditions'. The preconditions varied considerably.

2 Essential Background

2.1 In this section I set out in an abbreviated and very simplified form the background information essential to an understanding of the issues raised at the Inquiry, indeed to an understanding of the application itself and of the events which immediately preceded it. Most of such information is readily available in much fuller detail in published form but it appears to me essential that this report should be capable of being understood by a member of the public without knowledge of the nuclear power industry. This report is, as I understand it, intended to form, as was the Inquiry, an element in a wide public debate on nuclear issues. Moreover it was repeatedly stressed by one or other party in the course of the Inquiry that the public are badly informed and should be better informed. I have no doubt whatever that this is so, in the sense that the public should be provided with more in the way of digestible and reliable information. It is the lack of such information which renders the public or some members of it suspicious of those who operate the nuclear industry and exposes them to anxieties which are needless. In saying this I do not intend to imply that there are no grounds for anxiety in certain respects. There clearly are. It is equally clear, however, that many of the anxieties which are felt are without foundation and spring from a fear of anything nuclear, no doubt partly due to the fact that the Hiroshima and Nagasaki bombs with their devastating effects were the opening events in the development of nuclear power. Furthermore the anxieties which are felt, and deeply felt, however irrational and misplaced they may be, undoubtedly exist and are elements which must be taken into account.

Nuclear reactors and their fuel

2.2 The basic fuel source for commercial reactors both in the United Kingdom and elsewhere is uranium ore, from which natural uranium is extracted. The United Kingdom, so far as is presently known, has no appreciable indigenous supplies of uranium ore. Outside the communist countries the main deposits are to be found in the USA, Canada, Australia and South and Southwest Africa.

2.3 The atoms of uranium, like the atoms of many other elements, are of several different types. All these types of uranium (its 'isotopes') have identical chemical properties, but they do not behave in the same way. These different types of atom are distinguished by different numbers. Natural uranium consists for the most part of uranium 238 but it also comprises, to the extent of about 0.7 per cent, uranium 235. Uranium 235 atoms are fissile, that is to say they will, when irradiated with (struck by) neutrons, split or divide to form other lighter atoms, and at the same time discharge spare neutrons. In so doing they will release energy in the form of heat. The lighter atoms are known as fission products. Uranium 238 itself is not fissile but it will, if it absorbs or captures a spare neutron, as for example one of those discharged in a uranium 235 fission, rapidly undergo two sequences of radioactive decay to become plutonium 239. This is itself fissile. I deal with decay in paragraph 2.13 below.

2.4 In the United Kingdom two types of commercial reactor are currently in use, known respectively as Magnox reactors and Advanced Gas Cooled reactors (AGRs). Magnox reactors use natural uranium metal as their fuel. The metal, which is prepared in the form of rods, is encased or clad in a magnesium alloy known as magnox. Magnox reactors get their name merely from this circumstance.

2.5 AGRs require a fuel containing a higher proportion of uranium 235 than that present in natural uranium. This is achieved by adding to the uranium 235 present in a given amount of natural uranium further uranium 235, which has been extracted from additional amounts of natural uranium. This is the process known as enrichment. The result is enriched uranium. The extent of enrichment varies but the percentage of uranium 235 present in the fuel is increased from the natural level of 0.7 per cent to between 2 per cent and 3 per cent. The enriched uranium, prepared as uranium oxide, is formed into fuel pellets which are encased in a stainless steel tube. Stainless steel is used because it is necessary in order to be able to stand up to the much greater operating temperatures at which AGRs run.

2.6 It will be appreciated from the above that a tonne* of enriched uranium oxide fuel is much more costly than a tonne of natural uranium. Not only is more natural uranium required for its production, but extensive physical and chemical processes have to be employed to assemble the uranium 235 from, say, 5 tonnes of

---

*I use the metric tonne throughout for the sake of simplicity even where figures were expressed in evidence in tons. In such cases I have not converted to the equivalent in metric tonnes since, for present purposes, the difference is too small to be material.

natural uranium into a single tonne of enriched fuel. At the end of the enrichment process there will remain, at present without useful purpose, considerable quantities of uranium 238, together with a very small amount of uranium 235 which it would be too costly to extract. The more expensive process results in the fuel being able to stay in the reactor much longer and thus produce more heat and more electricity per tonne of fuel.

2.7 In the United Kingdom there are fourteen commercial nuclear power stations presently existing or under construction. Each of these stations generates its electricity from two nuclear reactors. Nine of the stations are powered by magnox reactors and five by AGRs. The five AGR stations are not yet all in operation but they will be in the near future. Two of the stations (one magnox and one AGR) are in Scotland. In England and Wales the net capability from nuclear stations as at 31 March 1977 was 3,462 megawatts (MW) out of a total capability of 56,365 MW, i.e. about 6 per cent. In Scotland, at the same date, the net capability from the two nuclear stations amounted to 1,300 MW, about 14 per cent of total generating capability. However, because it is the practice of the generating boards to provide base load electricity from nuclear stations, it is estimated that, when the stations at present under construction are completed, nuclear power will account for over 20 per cent of the electricity generated in England and Wales and about 30 per cent in Scotland. In addition there are certain further small (by present standards) nuclear power stations, which supply electricity to the grid. All commercial nuclear reactors are used, and can at present only be used, to generate electricity.

2.8 There are a number of other types of reactor currently in commercial use but of these it is only necessary, at this stage, to mention one, the Light Water Reactor (LWR). LWRs are not used in the United Kingdom but are relevant because, if the proposed development is permitted, it is intended by BNFL that LWR fuel should also be reprocessed. LWRs are also fuelled with enriched uranium oxide pellets but the extent of enrichment is slightly greater than in the case of AGRs and the tube in which the pellets are encased is made of an alloy known as zircaloy. As LWRs run at much lower temperatures than AGRs it is unnecessary to use stainless steel.

2.9 When the fuel is in the reactor, whether magnox, AGR or LWR, the quantity of uranium 235 is reduced by the fissions caused by collision with neutrons. At the same time plutonium 239 is created from uranium 238 in the manner described in paragraph 2.3 above. Some of this, too, is promply destroyed by fissions due to neutron collision but, when a fuel rod is removed from the reactor, a small amount will still remain in it.

2.10 For present purposes this matter may be summarised in this way: when a fuel rod leaves the reactor its contents will be:
  i. 97 per cent uranium (including what is left of the uranium 235 originally present – less than natural in the case of magnox fuel but still rather more than natural in the case of AGR and LWR fuel);
  ii. 0.1 per cent to 1 per cent plutonium;
  iii. 2–3 per cent of fission products and other radioactive substances formed by neutron absorption and subsequent decay in the same manner as has been described for the formation of plutonium 239. These latter fall within a group of substances composed of heavy atoms known as 'actinides'.

2.11 Finally, on the subject of reactors and their fuel, reference must be made to what is known as the Fast Breeder Reactor (FBR). None are at present in commercial use but several countries including the United Kingdom have built and are operating prototypes. The UK prototype is at Dounreay in Scotland. The FBR is so called because, by the use of the presently useless stocks of uranium 238 remaining after the enrichment process in combination with plutonium, more plutonium can be produced than is consumed by fissions which occur whilst the fuel is in the reactor. When, therefore, fuel comes out of the reactor, it may contain not only enough plutonium to provide its own replacement but more besides. This surplus can be accumulated until it is sufficient to charge an additional reactor.

2.12 There are a number of points which need to be noted about FBRs.
  i. Whilst they can be run so as to produce more plutonium than they consume they need not be so run, i.e. their introduction need not mean the production of ever increasing quantities of plutonium.
  ii. An FBR would, on current estimates, have to be run for some 25 years before it would produce enough plutonium for a further FBR.
  iii. The ability of FBRs to produce surplus plutonium does not mean that their introduction would make this country for ever independent of outside primary energy supplies. It does however mean that supplies would stretch very much further. In broad terms a stock of uranium that would fuel, say, 5 AGRs for 10 years, could fuel FBRs with a like generating capacity for a very much longer period, perhaps as much as 600 years.

Radioactivity and radiation

2.13 Radioactive substances are composed of atoms which are unstable, and which 'decay' with the discharge of particles or other radiation; and it is these radiations which may damage living tissues in various ways.

The principal types of radiation emitted during radioactive decay are known as alpha, beta and gamma radiation. Alpha radiation is only emitted by the atoms of heavy elements. It can only penetrate through water or body tissues by, at most, a few-hundredths of a millimetre. Beta radiation penetrates further through tissue, but never for more than a centimetre or two. Both alpha and beta radiation involve the discharge of particles. Gamma radiation involves electromagnetic rather than particulate radiation amd may penetrate through many tens of centimetres of tissue.

After any form of radioactive decay the original atoms change in their chemical property to that appropriate to the new atomic mass and charge which results from the particles they have lost in decay. The new atom may also be unstable and itself undergo radioactive decay. In this way a succession of atoms may be generated by the decay of one parent radioactive material, such 'decay chains' terminating only when a stable form of atom is formed as the final 'daughter product' of the chain.

2.14 The amount of damage caused by radiation will depend partly upon the particular organs or tissues which are irradiated, and partly upon the amount of energy delivered to these tissues by the radiation reaching them. In addition, certain types of radiation are more damaging, per unit of energy delivered, than others. Thus the alpha radiation given off in their decay by uranium, plutonium and other 'transuranic' elements (i.e. those with atoms heavier than that of uranium) is somewhat more damaging per unit of energy than the beta or gamma radiation given off by fission products. The dose (strictly the 'dose equivalent') resulting from a given radiation exposure is measured in a unit, the rem, which allows both for the amount of energy delivered to a tissue, and for the nature and hence the damaging effect of the radiations involved. Body tissues are normally exposed to about one-tenth of a rem (or 100 millirem) per year from natural sources of radiation.

This value varies, however, according to altitude, which affects the amount of radiation received from cosmic rays, and according to the nature of the soil or underlying rock. The influence of altitude is relatively small, the annual exposure being increased only by about 20 millirem (mrem) at a height of 5,000 ft above sea level. The background radiation in areas of granitic rock may however be raised by a rather greater amount. For example the annual exposure to a person living in Aberdeen, and occupying a granite house, would be about 200 mrem; and in some areas of the world where the underlying soil or sand contains substantial amounts of thorium, the exposure may be several times higher.

2.15 When a radioactive substance is released from a reprocessing plant, it is important to know for how long it will remain active and continue to emit radiation. This depends upon its half-period of radioactive decay, which states the time required for half of the atoms originally present to have undergone decay, and so for the radiation rate to have fallen to half its initial value. The activity of a given amount of radioactive substance is measured in curies or its sub-units, one curie corresponding to a particular number of atoms disintegrating per second.

2.16 If the body is exposed to radiation from a source outside itself, the external radiation dose received by any part of the body will depend upon the penetrating power of the radiation as well as upon the strength and position of the source. Alternatively, when the exposure is to internal radiation from radioactive substances that have been taken into the body, the dose to any part of the body will depend upon the organs in which the particular 'radionuclides' (i.e. radioactive substances) are concentrated and retained, as well as upon their half-period, their excretion rate, and the radiations which they emit.

2.17 Irradiation of body tissues may cause harm which is expressed either in the individual who is exposed to the radiation ('somatic' harm), or, as a result of damage to the germ cells, in the descendants (genetic harm). Except in the case of developmental defects resulting from pre-natal exposure, most somatic harm from low exposures is likely to consist of cancer or other malignancy developing in an organ or tissue which has been irradiated. Estimates have been made of the risk that such malignancy will develop following a given exposure of each of a number of body organs; and similar estimates have been made of the risk of genetic defects, per rem exposure of the germinal tissues.

2.18 To obtain some estimates, therefore, of the maximum risk of harm to any individual, or of the total risk to the population, from any particular release of radioactive materials into the environment, several steps are needed. It is first necessary, for each radionuclide discharged, to assess the ways in which the material, by virtue of its physical state or chemical form, may become distributed in the environment, and may enter the air that exposed persons may breathe or the food which they may eat. The maximum risk to any individual is thus derived from the amounts of each radionuclide that he may take into the body by breathing or through the mouth, and by irradiation from such materials present in his environment; and from the risks attributable to the resultant radiation exposure of his body or individual body tissues. Recommendations have been made as to the maximum annual radiation exposure that should be received by any individual, or by a so-called 'critical group' of individuals selected so as to be those who would receive the highest doses from the source under consideration.

2.19 In addition, it has been recognised as important to assess and review, not only the maximum risk of exposure or of harm that any individual might receive, but also the total exposure or harm for all individuals subject to irradiation from the particular source. For this purpose a 'collective dose' may be estimated, by multiplying the *average* dose received by all individuals exposed, by the number of individuals so exposed. The estimation of such a collective dose, in man-rem of total exposure, involves, in the case for example of atmospheric discharges, a determination of the distribution of wind direction and velocities, of population densities over the distances of importance in relation to the half-period of the radionuclides considered, and of the average respiratory and other body characteristics of the population. Similarly, for discharges into the sea, the concentration of different radionuclides from sea-water into marine species, and the total consumption of each relevant species, become important in estimating the collective doses due to dietary intake.

Reprocessing

2.20 The reprocessing problem is, in essence, the problem of deciding what should be done with fuel when it is removed from the reactors. In such a condition it is known as 'spent fuel'. At this stage it is highly radioactive and produces considerable heat. Before anything can be done with it, it must be stored for a period to allow the radioactivity and the heat to reduce to a point at which it may be handled under suitable shielding conditions without undue risk. This initial storage is done in cooling ponds, which provide both shielding and cooling, initially at the reactor sites and then, after transport from the reactor sites in special containers, at Windscale. I shall revert to the question of transport at a later stage.

2.21 Although the present application concerns oxide spent fuel from AGRs and LWRs, it is necessary first to consider the position with regard to magnox spent fuel because what has happened, is happening, and will continue to happen to such fuel is part of the essential background to BNFL's application.

Magnox spent fuel

2.22 The initial period in the cooling ponds at Windscale is necessarily comparatively short because the magnox cladding is subject to corrosion and, if it is breached, there will be an unacceptable escape of radioactivity to the pond water. When the cooling period has elapsed the fuel rods are removed, stripped of their cladding and subjected to a series of processes as a result of which:
  i. almost all the uranium is separated and stored for re-use in fuel for existing types of reactor;
  ii. almost all the plutonium is separated and stored for possible re-use, either in existing types of reactor or FBRs, should it be decided to introduce them;
  iii. the highly active fission products, the remaining traces of uranium and plutonium and the other actinides are combined in a single liquid stream which is then stored in shielded and cooled tanks known as Highly Active Waste Tanks (HAWs);
  iv. liquid, gaseous and solid low and medium active waste is created of which:
    a. low active liquid waste is discharged to the Irish Sea via pipe lines terminating 2.5 km off-shore and low active gaseous waste is discharged to the atmosphere via the plant stacks;
    b. low active solid waste is buried in trenches at BNFL's site at Drigg;
    c. medium active solid and liquid waste is stored for ultimate disposal at sea under international arrangements.

The above together comprise in very simplified form what is involved in reprocessing of magnox spent fuel.

2.23 BNFL's intention is that the highly active waste in the HAWs should ultimately be solidified by further processes into glass blocks, which blocks should thereafter be disposed of, either by burial deep in stable geological formations on land, or under the floor of the deep ocean or by deposit on the floor of the deep ocean. The process of conversion of the liquid waste to glass blocks is known as vitrification. The purpose of vitrification is to put the liquid waste into a form in which the radioactive substances which it contains can best be prevented from returning to the environment over the very long periods of time during which they will remain radioactive; periods which, for practical purposes, may be described as being 'for ever'. The vitrification process being developed by BNFL is known as HARVEST. Development began in the late 1950s but it was not until recently that a full research and development programme was put into operation. A full scale plant to demonstrate the process is expected to be in operation in the mid to late 1980s. Outline planning permission for such plant was granted on 1 March 1977.

2.24 Reprocessing of metal fuel has been carried out at Windscale since 1952, originally to separate plutonium for weapons use, and, in more recent years, in connection with the civil programme. Since 1964 reprocessing has been carried out in a building known as B.205. The operations have been generally successful save in two respects.
  a. Corrosion problems with the magnox cladding have resulted, since 1970, in greatly increased discharges of the fission products caesium 134 and 137 to the pond water and thence to the Irish Sea.
  b. There have also been difficulties in the first step in reprocessing, namely the stripping of the cladding, with the result that there has been a build-up of spent fuel in the ponds, which has itself aggravated the problem by increasing the pond storage time and thus the amount of corrosion.

In order to overcome these problems, which were increased by the fact that the provision of additional HAWs was delayed owing to the 3-day week, BNFL intend to provide new stripping facilities for magnox fuel and a pondwater treatment plant for the reduction of the caesium discharged to the Irish Sea to much lower levels. Outline planning permission for these two projects was granted by Cumbria County Council on 1 March 1977 (BNFL3).

2.25 It will be apparent from the above that BNFL have more than 25 years' experience of reprocessing metal fuel. Since they began to do so they have reprocessed some 19,000 tonnes. In the course of so doing they have separated and stored for possible future use some

10 tonnes of plutonium of which some 7½ tonnes remain in store. Continued reprocessing of magnox fuel is estimated to yield a further 45 tonnes of plutonium by about the year 2000. Thus, whatever is done with regard to oxide spent fuel we shall, by the year 2000, have separated 52½ tonnes of plutonium from magnox fuel. We shall also have in store in HAWs large amounts of highly active waste if none of such waste has by then been vitrified.

2.26 It is clear that the vitrification process must be developed for the magnox highly active liquid waste. Having proceeded to the stage of such waste there is no practical alternative. I have, on the evidence, little doubt that it will be developed successfully. Both this matter and the question whether, when vitrified, the blocks can be safely disposed of I shall consider at a later stage.

Oxide spent fuel

2.27 The reprocessing of oxide spent fuel has, in general, the same results as the reprocessing of magnox fuel, although, for a given tonnage of oxide fuel, the radio-active content of the spent fuel is some ten times higher. If permission to develop is granted and if BNFL go

Environmental impact of discharges from Windscale expressed as percentage of present maximum permitted radiation doses

Radio-nuclide	Critical pathway	1975 (Magnox plant)	Intentions for new plants		
			Refurbished magnox	THORP excluding (including) margins	Refurbished magnox plus THORP excluding (including) margins
Aqueous discharges					
Cs134/137	Fish	24	2.75	0.22 (1.1)	3.0 (3.9)
Sr90	Fish	2.0	1.1	0.05 (0.25)	1.2 (1.4)
H3	Fish	0.004	0.005	0.1 (0.1)	0.1 (0.1)
I.129	Fish	0.04	0.06	0.5 (0.5)	0.6 (0.6)
	Total			0.87 (1.95)	4.9 (6.0)
Ru106	Silt	1.9	0.78	0.14 (0.7)	0.9 (1.5)
Zr95/Nb95	Silt	1.5	0.11	0.05 (0.25)	0.2 (0.4)
	Total			0.19 (0.95)	1.1 (1.9)
Alpha	Resuspension from silt (Inhalation)	0.13	0.04	0.02 (0.1)	0.06 (0.1)
Atmospheric discharges					
Kr	Immersion (skin)	0.08	0.11	0.9 (0.9)	1.0 (1.0)
H3	Inhalation and food	0.04	0.06	0.04 (0.2)	0.1 (0.3)
C14	Inhalation and food	0.06	0.09	0.15 (0.15)	0.2 (0.2)
I.129	Milk	0.2	0.8	0.15 (0.6)	1.0 (1.4)
Ru106	Inhalation	0.14	0.14	0.05 (0.5)	0.2 (0.6)
Sr90	Milk	1.4	0.58	0.36 (3.6)	0.9 (4.2)
Alpha emitters	Inhalation	2.0	0.2	0.25 (2.5)	0.5 (2.7)
	Total			1.9 (8.45)	3.9 (10.4)
Totals all pathways		33.49	6.825	2.98 (11.45)	9.96 (18.4)

Notes
1. The figures for refurbished magnox are based on (i) a throughput of 1,500 tonnes uranium per annum, which is higher than ever before and the highest at which the plant is likely to operate, (ii) a cooling period of one year and (iii) the longest achievable time in the reactor – 3,500 megawatt days (MWD) per tonne uranium. The figures therefore cover the worst possible situation.
2. The figures for THORP are based on (i) a throughput of 1,200 tonnes uranium per annum i.e. the maximum theoretical capacity (ii) LWR fuel one year cooled – an abnormally short period (iii) the longest achievable time in the reactor – 37,000 MWD per tonne uranium. Again, therefore, the figures cover the worst possible situation.
3. The 1975 figure for H3 (tritium) is an assessment based on the measured discharge of krypton 85.
4. The 1975 figure for carbon 14 is calculated.

ahead as planned there will be:
   a. more plutonium and uranium separated;
   b. more discharges to the sea and the atmosphere, more material for burial at Drigg, more dumping in the deep oceans and more storage of highly active waste.

2.28 The additional plutonium separated, on the assumption that the plant is run at 50 per cent designed capacity for 10 years, would be about 40 tonnes and the volume of highly active waste produced would be about 30,000 cubic metres.

So far as discharges to sea and atmosphere are concerned I set out above in tabular form BNFL's estimates of the environmental impact of (a) the present discharges, (b) the discharges after the new magnox facilities, for which outline consent has been granted, have been completed, (refurbished magnox) (c) the discharges from both refurbished magnox and THORP. The expressions 'excluding margins' and 'including margins' which appear in the table require explanation. 'Excluding margins' indicates that the figures are BNFL's estimates of what the impact will be. The project is however, at present, only at an early stage of design. BNFL recognise that, as the project proceeds, variations may occur. They have therefore provided a separate set of figures to cover the maximum upward variations which they consider might occur when and if the plant finally went into operation. The phrase 'including margins' is used to describe figures arrived at on this basis.

2.29 It will be seen from the above table that, if THORP is built, the totals of the discharges from Windscale (even including margins) are expected to result in radiation exposures representing small fractions only of maximum permitted doses. The risks involved in exposure to radiation are discussed in paragraphs 10.26 to 10.36 below, but it is convenient at this stage to give a comparison which will serve to indicate the general level of radiation under consideration. I have already mentioned (para 2.14) that everyone normally receives a radiation dose to body tissues of about 100 mrem per annum. The present limit (apart from medical) for whole body exposure of members of the general public, is 500 mrem per annum so that, if exposure were kept to 10 per cent of the limit, the dose received would be no more than 50 mrem per annum. This is about half as much as the additional natural radiation that would be suffered by a person who moved from a brick house in London to a granite house in Aberdeen. I do not, in giving this example, mean to suggest in any way that exposure to radiation at 10 per cent of the limit can be ignored. It is common ground that all additional radiation is harmful to some extent. But there is no difference between natural radiation and man-made radiation. It is therefore, relevant to know that, if BNFL fulfil their expectations, the extent of harm from radiation is likely to be considerably less than that which would be suffered as a result of a move in residence from one place to another in the United Kingdom.

Oxide fuel arisings to the year 2000

2.30 Those AGRs already in operation or under construction in the UK will have given rise by 1995 to some 3,150 tonnes of spent fuel which must be dealt with in some way or other. Spent fuel will continue to be produced thereafter from those AGRs at the rate of about 200 tonnes per annum. Hence, by the year 2000, accumulations from existing reactors will be 4,150 tonnes if no reprocessing takes place, to which must be added a further 200 tonnes from prototype reactors operating at Windscale and Winfrith. If, as appears likely, reactors to produce a further 4,000 MW per year of electricity are ordered in the near future and begin to operate between 1990 and 1995 they will, by the year 2000, have produced about a further 1,700 tonnes of spent fuel. Thus a total of 6,000 tonnes by the year 2000 from UK reactors alone is a realistic forecast.

2.31 At the same time world arisings of spent fuel will continue to be created. If not reprocessed, these arisings, with their content of fission products, plutonium and other actinides, will continue to accumulate. An idea of the scale of the arisings can be gained from the fact that arisings by 1990 from reactors already existing or under construction in non-communist countries, excluding France, UK and USA, are estimated by BNFL to amount to about 20,000 tonnes. This estimate was not seriously challenged and I accept it. If USA and French arisings are added the figure must at least double. It is necessary to keep in mind the world situation for, if radioactivity escapes to the environment, it can reach any part of the world.

2.32 If spent fuel arisings are not reprocessed as presently contemplated the alternatives appear to be:
   i. medium or long-term storage of the fuel elements followed by final disposal of the fuel elements as such;
   ii. permanent storage of the fuel elements;
   iii. medium or long-term storage, followed by a form of reprocessing (which would separate the uranium but not the plutonium) followed by final disposal of the residue in some form or another.

I revert to these matters later.

BNFL's capability without THORP

2.33 Between 1969 and 1973, BNFL in fact reprocessed 100 tonnes of oxide spent fuel. Building B204 was used for the initial stages of reprocessing and B205, when it was not being used for magnox reprocessing, for the later stages. In 1973, when further reprocessing was about to be undertaken, an accident occurred which resulted in B204 being shut down. It has been shut down ever since. BNFL intend to carry out modifications to it, which would enable it to be used again for the initial stages of oxide reprocessing, the final stages being done in B205. These modifications could be completed in 1979 and from then until 1984, when the new magnox facilities should be finished, there would be sufficient spare capacity in B205 to clear all accumulated UK arisings. Thereafter from 1981–1984 there would be no spare capacity in B205 for it would be in use at full capacity to reprocess the accumulated backlog of magnox fuel. From 1984 onwards there would, however, again be spare capacity in B205. This should be more than sufficient to deal with any oxide fuel which required to be reprocessed owing to deterioration of its cladding in storage. It should also be sufficient, in theory, to handle arisings from UK reactors presently existing or under construction. It could do so provided that nothing went wrong, but it would leave no reasonable margin for untoward occurrences. BNFL and the home generating boards consider this to be an unacceptable means of reprocessing home arisings, if they are to be reprocessed. I accept this contention, not only on the basis of commercial good management but also because B204 is old. It must, I consider, be less safe for its operators than a new plant, particularly if it was being worked to capacity.

Its proper use is, in my view, for the reprocessing of spent fuel, the condition of which makes reprocessing at any particular time necessary, and for gaining further experience of reprocessing of oxide fuel. Such experience would be of value whether or not BNFL's present application succeeds.

I should add that the B204/205 reprocessing route would not, even in theory, be capable of handling arisings from the additional reactors which seem likely to be ordered or the 1,150 tonnes of foreign fuel, to the reprocessing of which BNFL are already committed.

Plutonium

2.34 Certain facts about plutonium are given here because it was apparent to me that there exists much misunderstanding about it.
1. It is not true that plutonium never existed until man made it. It was stated on behalf of one party at an early stage in the case that God never made plutonium. Later, that party's own expert witness accepted that the existence of a natural nuclear reactor, which had made plutonium in the long distant past in Gabon had been established. To talk of the creation of plutonium as 'man's bargain with the Devil' or 'the Faustian bargain' is therefore no more than emotive nonsense.
2. It is not true that plutonium is highly radioactive. Its principal isotope plutonium 239 is relatively stable and as a consequence its half-life is very long and its radioactivity (per unit mass) is very low.
3. It is not true that plutonium has only two uses, making bombs and making electricity commercially. Plutonium 238 is used within the body as the power source for heart pacemakers.
4. It is not true that in all circumstances very small amounts of plutonium are lethal. Insoluble particles when inhaled certainly are hazardous in small quantities. Considerably larger amounts could be eaten without appreciable harm.
5. It is not true that plutonium is only safe when protected by massive shielding. As regards shielding from its radiation, it could be sat on safely by a person with no greater protection than, as Professor Fremlin put it, ' a stout pair of jeans'.
6. It is not true that plutonium is the most toxic substance known to man. Numerous radionuclides are more toxic than plutonium 239 if present in food or water, and particularly the isotopes of radium, two of which are over 100 times as toxic when the comparison is made between soluble forms. Similarly, several of the isotopes of thorium are rather more toxic than plutonium 239 if inhaled, if one compares insoluble forms.
7. It is not true that an escape of plutonium would be a unique disaster. The damage done, for example, by the breaking open of a tanker of chlorine of the size which regularly travels by road and rail would be a great deal more damaging than the breaking open of a container of spent fuel with its plutonium content.

On the other hand it is true
1. That plutonium is a bomb-making material.
2. That if plutonium reaches a critical mass there will be a chain reaction and thereby the creation of highly active fission products.
3. That in certain circumstances plutonium is very dangerous to man.
4. That plutonium, if released into the environment, persists for a very long time.
5. That, as a result, stringent precautions are necessary to prevent plutonium falling into the wrong hands, from reaching critical masses and from returning to man over the long period of its life.
6. That, as was readily accepted by Friends of the Earth Ltd (FOE), it is in everyone's interest to find as safe as possible a resting place for atomic waste, whether in the form of spent fuel containing plutonium or in the form of glass blocks containing only about $\frac{1}{1,000}$ of the amount of plutonium that would be contained in the spent fuel. There was, rightly, much stress laid upon our obligations to future generations. These obligations include the obligation to find a safe resting place for our waste if we can, rather than leave it for them to do so. Resistance to such attempts is neither in their interests nor in our own. In whatever form the waste is to be, it is likely to be safer in deep holes in stable geological formations than preserved in above ground storage.

3 History of the Application

3.1 From the time when nuclear power was first introduced until comparatively recently, it was generally assumed that spent fuel would be reprocessed and the uranium and plutonium extracted. The uranium would be used in the types of reactor which are now in use, known generally as 'thermal' reactors, and the plutonium would be used in FBRs. Indeed in his book 'Nuclear Power' published in 1976 (FOE9), Walter C. Patterson, who was the leading witness called on behalf of FOE, (one of the principal objectors) wrote (p 100):

> 'Fissile uranium and plutonium are much too valuable to be thrown away. Even if it were not valuable, plutonium is in any case too dangerous to be let loose in the environment. Nor must the remains of the fuel, including the fission products, be thrown away – not because of their value but because they too are dangerously radioactive. Accordingly, the irradiated fuel from a reactor is usually "reprocessed".'

In evidence Mr Patterson stated that although the book was published in 1976 the passage which I have quoted was written in 1974 and that he had since changed his opinion. I quote it simply to indicate that opposition to reprocessing, even amongst some of those most wary of the implications of nuclear power, is of recent origin.

3.2 It was in this general situation that BNFL, in late 1974, first announced its expansion plans which included the building of THORP. During 1975 and the first half of 1976 discussions were held with Cumbria, Copeland and other interested authorities. In addition there were meetings to discuss, in particular, the question of reprocessing foreign fuel. This phase in the history came to an end when, on 12 March 1976, the Secretary of State for Energy announced that the Government had decided that BNFL might take on further reprocessing work for overseas customers, subject to the negotiation of satisfactory terms, including the option to return radioactive waste.

3.3 Thereafter, on 1 June 1976, BNFL made an application for outline planning permission for their entire expansion plans. This was, like the present application, referred to Cumbria as a county matter. There followed a number of discussions, including a public meeting at Whitehaven on 29 September 1976, and, on 2 November 1976, Cumbria's Town and Country Planning Committee, although minded to approve the application, resolved to refer it on the basis that it involved a departure from a fundamental provision of the County Development Plan. Whether the Committee were right or wrong in taking the view that the entire development would constitute such a fundamental departure is of no importance in the light of the subsequent events. On 22 December 1976 you announced that you considered that that part of the proposal relating to the oxide reprocessing plant should be called in so that you might satisfy yourself that the proposal was acceptable. As a result of this announcement BNFL, by letter dated 21 January 1977, withdrew the part of the proposal relating to such plant and on 1 March 1977 outline planning permission was, as previously mentioned, granted for the remainder. On that same day BNFL submitted four separate applications as follows:

    a. for the whole plant excluding the fuel receipt and storage facilities;
    b. for the whole of the receipt and storage facilities;
    c. for the first phase only of the receipt and storage facilities.

All the above were for outline permission.

    d. for the extension of an existing oxide storage fuel pond in order to enable fuel from already existing overseas contracts to be received. This application was for full planning permission.

Application (a), the present application, was, as previously mentioned, called in. Application (b) was deferred. Application (c), which was designed to enable the possibility of foreign business then under negotiation to be kept open pending a decision on the present application, was granted on the basis that any fuel received would be returned in the event of the present application being refused. Application (d) was granted.

4 Scope of the Inquiry

4.1 As is already generally known the Inquiry was very wide-ranging and many of the matters raised may not have appeared to be of particular relevance to the question in hand, namely whether outline consent should be given for an oxide reprocessing plant at Windscale of an annual nominal capacity of 1,200 tonnes to reprocess both UK and foreign spent fuel. The terms of the Rule 6(1) Statement appeared to me however to entail a very wide investigation. Indeed, quite apart from such statement, such an investigation was, in the prevailing circumstances, desirable. Nevertheless it is necessary to remember that the application under scrutiny is for outline planning permission only and that, even if consent is granted, THORP might never be built. This might happen for any one of a number of reasons, e.g. because this country entered into an international agreement not to reprocess; or because BNFL could not obtain sufficient firm commitments for reprocessing to make the project, in their view, financially sound; or because they could not obtain the necessary permission from the control authorities.

4.2 The Inquiry was also wider than it might otherwise have been owing to the fact that there are a number of policy decisions with regard to the future which have yet to be taken and that, as a result of the Government Response (Cmnd 6820) (BNFL 170) to the Sixth Report of the Royal Commission on Environmental Pollution (Cmnd 6618) (the Sixth Report) (BNFL 9) parts of the existing structure of control are to be changed or are under review.

I shall hereafter refer to the Government Response as 'The White Paper'.

5 Summary of Contentions and Structure of the Report

Applicant's case

5.1 BNFL's case at the opening of the Inquiry may be summarised as follows:
1. BNFL has the necessary technical experience to develop and operate the proposed plant. Magnox fuel has been reprocessed successfully for 25 years. There is also experience of separating and storing plutonium and of reprocessing oxide fuel.
2. Reprocessing is desirable as an energy conservation measure and would add to secure indigenous fuel resources.
3. Whilst the proposed development is not dependent on a decision whether or not to go ahead with FBRs, it is essential if the FBR option is to be kept open.
4. Reprocessing of spent fuel from UK AGRs in operation or already under construction is essential on waste management grounds.
5. Existing plant in the UK is inadequate to deal with anticipated AGR spent fuel arisings.
6. A plant large enough to reprocess foreign spent fuel in addition to UK arisings would permit economies of scale and would bring a balance of payments advantage to the UK.
7. Foreign business exists which would justify construction of a plant of 1,200 tonnes capacity.
8. The UK reprocesses fuel for foreign customers under internationally accepted safeguards designed to prevent the proliferation of nuclear weapons. If we were to deny reprocessing services to foreign customers they might be driven to develop their own facilities without the protection the safeguards provide. Such denial might therefore add to the risks of proliferation.
9. Terrorism will continue to find targets and to present a threat whether or not reprocessing and plutonium separation continue on an increased scale at Windscale. The additional risk and threat to the civil liberties posed by the proposed development would therefore be negligible.
10. Reprocessing technology is not novel and the Nuclear Installations Inspectorate (NII) are confident that the proposed plant can be designed, built and operated to high standards of safety.
11. The emissions of radioactivity from the plant during routine operation give no grounds for supposing that employees or the public at large will face any significant risk.
12. The effect of the plant on visual amenity and infrastructure raises no problems which cannot be satisfactorily resolved.
13. The development would create a substantial number of stable jobs in a Special Development Area with a higher than average level of unemployment.

At the end of the Inquiry their case remained substantially the same.

Objectors' cases

5.2 In broad terms the objections raised by the various objectors were to the following effect:
1. The plant would increase the dangers of nuclear weapons proliferation.
2. The plant would create unacceptable risks from terrorism. Alternatively, the containment of such risks within acceptable levels could only be achieved at the cost of an interference with civil liberties which would itself be unacceptable.
3. There is in any event no present need for the plant and will probably never be such a need.
4. Permission would pre-empt a decision on the Fast Breeder question.
5. The plant would be an unsound proposition on financial grounds.
6. Emissions from the plant in normal operation would create unacceptable risks to the workforce, to the public, to future generations, and to the natural environment.
7. The risks to the workforce from minor incidents at the plant and, to the public and the environment, from major accidents at the plant would be unacceptable.
8. The risks to transport workers and the public from accidents in the course of transporting spent fuel to the plant, or fresh fuel or plutonium from the plant, would be unacceptable.
9. It is not yet established that the highly active waste resulting from reprocessing can be safely disposed of by means of vitrification and burial of the resulting glass blocks. Disposal of spent fuel as such, without reprocessing, might prove preferable and no further highly active waste should be created until this possibility has been fully researched and the position established one way or the other. Other methods may also be found.

10. In any event foreign fuel should not be reprocessed.
11. Even if the intended limits of radioactive discharges and the estimated accident risks were acceptable, the plant would represent too ambitious an advance in technology and there could be no confidence that the plant would operate so as to confine the discharges and risks as intended and estimated.
12. The presently prevailing institutional arrangements for fixing limits of radiation doses and discharges, for vetting the design, construction and operation of plants producing radioactive emissions and for monitoring discharges from such plants give no grounds for confidence that the various authorities are sufficiently independent or competent to protect the public.
13. There is emotional hostility to the project in a large section of the public which could lead to violence and permission should be refused on this ground alone.
14. If there is to be an oxide reprocessing plant in the UK it has not been established that Windscale is the proper location for it, indeed it is a bad location.
15. Although the plant would create new jobs in West Cumbria, which is an area of high unemployment, the number of such jobs likely to be filled by the unemployed would be relatively few. A large number of the available jobs would go to immigrants into the area, whose arrival would impose severe strains on housing, sewerage, roads and the like.
16. The additional jobs are, in any event, not of the most desirable nature because they would be provided by a company which is already a dominant employer in the area.
17. The nature of the Inquiry, the interval between its arrangement and opening, the lack of adequate information preceding it and the disparity of the resources available respectively to the applicants and the objectors has resulted in an inadequate investigation of the issues. There is therefore no satisfactory basis for a decision in favour of the applicant.

5.3 To attempt hereafter to deal separately and consecutively with each of the numbered contentions of the parties would lead to a great deal of overlap and confusion. I have therefore sought, in the remainder of the report, to cover the various contentions under main headings which will, so far as possible, avoid overlapping and enable the issues to be considered in some reasonably logical sequence.

5.4 I have in a number of respects departed from the normal format of a local planning inquiry report. The nature of the Inquiry itself has rendered this inevitable. It is only necessary to make specific mention of one such departure. That is the omission of any details of the site and its surroundings or of the project itself. To have included such details would have added greatly to the length of what is, in any event, a very long report and would not have served any useful purpose. I have mentioned therefore only such of the details as are immediately relevant to some particular issue raised at the Inquiry. Should it be desired to consider all the details, whether or not relevant to an issue specifically raised, they may be found, as to the site and its surroundings, in a statement of facts agreed between the applicants, Cumbria and Copeland (CCC34) and, as to the project itself, in the proof of evidence of Mr B. F. Warner and the Appendices thereto.

6 The Nuclear Weapons Proliferation Question

6.1 The possible effect of the building of THORP upon the spread of nuclear weapon capability was much canvassed before me. It formed the main ground upon which FOE submitted that a decision on the building of the plant should be delayed for at least ten years and thus that the present application should be rejected. In this they were supported by a number of other objectors and those who did not positively object on this particular ground expressed anxieties in connection with it. BNFL on the other hand contended that the building of THORP, far from tending to increase or accelerate the spread of nuclear weapon capability, would tend to decrease or delay such spread.

6.2 A nuclear bomb can be constructed with the grade of plutonium recovered by reprocessing. A country, which had in its hands such plutonium, could produce a bomb or bombs more rapidly, and with less risk of its actions being detected in time for international diplomatic pressure to be exerted, than if it had no such plutonium. It was submitted, therefore, that if THORP were built and used to reprocess foreign fuels, and if the recovered plutonium were returned to the countries concerned, this must inevitably increase the proliferation risks. This argument does not apply to the reprocessing of UK fuel, both because we already have a nuclear weapon capability and because the plutonium already recovered and yet to be recovered from magnox fuel is enough to manufacture a great number of bombs. Nor does the argument apply to the reprocessing of fuel from, and return of the recovered plutonium to, countries which, like ourselves, are already nuclear weapon powers. It is, however, contended that, even if THORP were used wholly for the reprocessing of fuel from UK reactors and from nuclear weapon powers, it would still indirectly increase the risk of proliferation on the grounds:
  a. that the plutonium might be stolen whilst in transport;
  b. that, if the UK were to embark on reprocessing, it would be difficult if not impossible to prevent other countries also doing so, with the result that they would then be in a position to move rapidly to the creation of nuclear weapons.

6.3 The contrary argument is (a) that the reprocessing of foreign fuel would lessen the incentive of the countries sending fuel for reprocessing to develop their own reprocessing facilities and (b) that, if the plutonium were returned in the form of fuel rods, after brief irradiation to make them dangerous to handle, this would both practically eliminate the risks of theft in transport and render reprocessing of the irradiated fuel rods necessary before weapon material would be available. This would, it was argued, be preferable to driving other countries into developing their own reprocessing facilities.

6.4 The contention that THORP would have a proliferating effect was supported by both oral and documentary evidence from a number of eminent people with wide knowledge of the problems involved. The principal witnesses who gave evidence on this subject were: for FOE, Mr Walter Patterson and Professor A. Wohlstetter, Professor of Political Science at the University of Chicago; for the Natural Resources Defence Council (NRDC), Dr T. B. Cochran, Physicist Staff Scientist; for the National Peace Council (NPC), Mrs Sheila Oakes and for the Town and Country Planning Association (TCPA), Professor J. Rotblat, Emeritus Professor of Physics at the University of London.
   The opposite view was supported by the oral evidence of Dr D. G. Avery of BNFL and Mr C. Herzig from the Department of Energy. It, too, was also supported by documentary evidence.

6.5 An evaluation of the opposing contentions requires an examination of the facts of, and leading up to, the present situation. It is first necessary to observe that the supply of plutonium to non-weapon countries has been going on for a considerable time, as has the supply of uranium enriched to more than 20 per cent in uranium 235, at or above which level of enrichment it is regarded as weapon material, and of uranium 233 which also is fissile material. BNFL has itself exported plutonium to a number of such countries under written Government authorisations. The USA has exported considerable quantities of all three substances. I had no evidence before me of the accumulated total exports from the USA up to the present time but Professor Wohlstetter, in an article entitled 'Spreading the Bomb Without Quite Breaking the Rules' (FOE28), written and produced in evidence by him, stated:

> 'We (the USA) have for some time exported to non-weapon states, for use in research, both separated plutonium and highly enriched uranium, which bring them closer to the bomb than do the facilities for separating such material. For example, from mid-1968 to spring 1976 we exported 697 kilograms of highly enriched uranium and 104 kilograms of separated

plutonium to Japan and 2,170 kilograms of highly enriched uranium and 349 kilograms of separated plutonium to the Federal Republic of Germany.' Furthermore a table contained in the Pelican book 'Soft Energy Paths' by Amory B. Lovins (WA 150) sets out total gross Exports of Strategic Nuclear Materials from the USA up to 31 March 1976. This shows supplies to a large number of non-nuclear-weapon countries in more than sufficient quantities to make one or more bombs. Such exports have been made under contracts containing undertakings to use for peaceful purposes and to accept the application of various safeguards. These undertakings, so far as is known to me, have been honoured.

6.6 At present the system for preventing the spread of nuclear weapons is founded on a number of agreements of which the principal ones are the 1956 International Atomic Energy Agency (IAEA) Statute (BNFL269), the 1957 Treaty Establishing the European Atomic Energy Community (EURATOM) (BNFL50) and the 1970 Treaty on the Non Proliferation of Nuclear Weapons (NPT) (BNFL51). It is necessary to refer to certain provisions of those three documents. Before doing so, however, it should be mentioned that the system of safeguards which they contain or for which they provide is essentially one of reporting and inspection. This system was acknowledged by everyone to be in need of strengthening and improvement. I shall not therefore lengthen this report by describing the system and its shortcomings. It is sufficient to say that it could and should be improved, e.g. by increasing the numbers of inspectors and, possibly, by the development and introduction of improved methods for detecting any diversion of fissile material from peaceful uses.

6.7 The IAEA Statute contains the following immediately relevant provisions:

'**Article II – Objectives**
The Agency shall seek to accelerate and enlarge the contribution of atomic energy to peace, health and prosperity throughout the world. It shall ensure, so far as it is able, that assistance provided by it or at its request or under its supervision or control is not used in such a way as to further any military purpose.
Article III – Functions
1. To encourage and assist research on, and development and practical application of atomic energy for peaceful uses throughout the world; and, if requested to do so, to act as an intermediary for the purposes of securing the performance of services or the supplying of materials, equipment, or facilities by one member of the Agency for another; and to perform any operation or service useful in research on, or development of practical application of, atomic energy for peaceful purposes.
2. To make provision, in accordance with this Statute, for materials, services, equipment, and facilities to meet the needs of research on, and development and practical application of, atomic energy for peaceful purposes, including the production of electrical power, with due consideration for the needs of the under-developed areas of the world.
3. To foster the exchange of scientific and technical information on peaceful uses of atomic energy.
4. To encourage the exchange and training of scientists and experts in the field of peaceful uses of atomic energy.
5. To establish and administer safeguards designed to ensure that special fissionable and other materials, services, equipment, facilities, and information made available by the Agency or at its request or under its supervision or control are not used in such a way as to further any military purpose; and to apply safeguards, at the request of the parties, to any bilateral or multilateral arrangement, or at the request of a State, to any of that State's activities in the field of atomic energy.
Article XX – Definitions
1. The term 'special fissionable material' means plutonium 239; uranium 233; uranium enriched in the isotopes 235 or 233; any material containing one or more of the foregoing; and such other fissionable material as the Board of Governors shall from time to time determine; but the term 'special fissionable material' does not include source material.
3. The term 'source material' means uranium containing the mixture of isotopes occuring in nature; uranium depleted in the isotope 235; thorium; any of the foregoing in the form of metal, alloy, chemical compound, or concentrate; any other material containing one or more of the foregoing in such concentration as the Board of Governors shall from time to time determine; and such other materials as the Board of Governors shall from time to time determine'.

At the time when the Statute was entered into it was generally accepted that the future of nuclear power included the use of plutonium 239 in FBRs. The provisions quoted above are sufficient to show that the intention then was that plutonium 239 should be separated, that the technology both for reprocessing and for FBRs should be developed and freely exchanged, and that plutonium 239 should be made available to members.

6.8 The same intention can be seen in the EURATOM Treaty. I refer simply to:

'**Article 52**
1. The supply of ores, source materials and special fissile materials shall be ensured, in accordance with the provisions of this Chapter, by means of a common supply policy on the principle of equal access to sources of supply;
and
Article 93
Member States shall abolish between themselves, one year after entry into force of this Treaty, all customs duties on imports and exports or charges having equivalent effect, and all quantitative restrictions on imports and exports, in respect of:
  a. products in Lists A^1 and A^2;'

List A¹ includes 'uranium enriched in uranium 235', 'uranium enriched in plutonium' and 'plutonium' itself.
List A² includes:

'Equipment specially designed for the chemical processing of radioactive material:
- equipment for the separation of irradiated fuel;
- by chemical processes (solvents, precipitation, ion exchange, etc);
- by physical processes (fractional distillation etc);
- waste processing equipment;
- fuel recycling equipment.'

The intention is made very clear by the specific provision for the inclusion of both plutonium and reprocessing equipment in a Nuclear Common Market.

6.9 I come now to the NPT itself, to which there are at present 103 parties. Again, at the time it was entered into, it was generally accepted that the future of nuclear power lay in reprocessing and the use of separated plutonium in FBRs.

6.10 The preamble to the NPT includes the following:—
'The States concluding this Treaty ...

Undertaking to co-operate in facilitating the application of International Atomic Energy Agency safeguards on peaceful nuclear activities.

Expressing their support for research, development and other efforts to further the application, within the framework of the International Atomic Energy Agency safeguards system, of the principle of safeguarding effectively the flow of source and special fissionable materials by use of instruments and other techniques at certain strategic points.

Affirming the principle that the benefits of peaceful applications of nuclear technology, including any technological by-products which may be derived by nuclear-weapon States from the development of nuclear explosive devices, should be available for peaceful purposes to all Parties to the Treaty, whether nuclear-weapon or non-nuclear-weapon States.

Convinced that, in furtherance of this principle, all Parties to the Treaty are entitled to participate in the fullest possible exchange of scientific information for, and to contribute alone or in co-operation with other States to, the further development of the applications of atomic energy for peaceful purposes.'

6.11 The expression 'source and special fissionable materials' is not defined in the Treaty but, in view of the reference to the IAEA safeguards, there can be little doubt that it was to have the same meaning as in the IAEA statute. The reference to safeguarding the flow of special fissionable material must therefore be read as including the safeguarding of the flow of plutonium.

6.12 Immediately relevant provisions of the Treaty itself are:—

'**Article I**

Each nuclear-weapon State Party to the Treaty undertakes not to transfer to any recipient whatsoever nuclear weapons or other nuclear explosive devices or control over such weapons or explosive devices directly, or indirectly; and not in any way to assist, encourage, or induce any non-nuclear-weapon State to manufacture or otherwise acquire nuclear weapons or other nuclear explosive devices, or control over such weapons or explosive devices.

Article II

Each non-nuclear-weapon State Party to the Treaty undertakes not to receive the transfer from any transferor whatsoever of nuclear weapons or other nuclear explosive devices or of control over such weapons or explosive devices directly, or indirectly; not to manufacture or otherwise acquire nuclear weapons or other nuclear explosive devices; and not to seek or receive any assistance in the manufacture of nuclear weapons or other nuclear explosive devices.

Article III

1. Each non-nuclear-weapon State Party to the Treaty undertakes to accept safeguards ... for the exclusive purpose of verification of the fulfilment of its obligations assumed under this Treaty with a view to preventing diversion of nuclear energy from peaceful uses to nuclear weapons or other nuclear explosive devices. Procedures for the safeguards required by this Article shall be followed with respect to source or special fissionable material whether it is being produced, processed or used in any principal nuclear facility or is outside any such facility ...
2. Each State Party to the Treaty undertakes not to provide: (a) source or special fissionable material, or (b) equipment or material specially designed or prepared for the processing, use or production of special fissionable material to any non-nuclear-weapon State for peaceful purposes, unless the source or special fissionable material shall be subject to the safeguards required by this Article.
3. The safeguards required by this Article shall be implemented in a manner designed to comply with Article IV of this Treaty, and to avoid hampering the economic or technological development of the Parties or international co-operation in the field of peaceful nuclear activities, including the international exchange of nuclear material and equipment for the processing, use or production of nuclear material for peaceful purposes in accordance with the provisions of this Article and the principle of safeguarding set forth in the Preamble of the Treaty.
4. Non-nuclear-weapon States Party to the Treaty shall conclude agreements with the International Atomic Energy Agency to meet the requirements of this Article either individually or together with other States in accordance with the Statute of the International Atomic Energy Agency. Negotiation of

such agreements shall commence within 180 days from the original entry into force of this Treaty. For States depositing their instruments of ratification or accession after the 180-day period, negotiation of such agreements shall commence not later than the date of such deposit. Such agreements shall enter into force not later than 18 months after the date of initiation of negotiations.

Article IV

1. Nothing in this Treaty shall be interpreted as affecting the inalienable right of all the Parties to the Treaty to develop research, production and use of nuclear energy for peaceful purposes without discrimination and in conformity with Articles I and II of this Treaty.

2. All the Parties to the Treaty undertake to facilitate, and have the right to participate in, the fullest possible exchange of equipment, materials and scientific and technological information for the peaceful uses of nuclear energy. Parties to the Treaty in a position to do so shall also co-operate in contributing alone or together with other States or international organisations to the further development of the applications of nuclear energy for peaceful purposes, especially in the territories of non-nuclear-weapon States Party to the Treaty, with due consideration for the needs of the developing areas of the world.

Article X

1. Each Party shall in exercising its national sovereignty have the right to withdraw from the Treaty if it decides that extraordinary events, related to the subject matter of this Treaty, have jeopardised the supreme interests of its country. It shall give notice of such withdrawal to all other Parties to the Treaty and to the United Nations Security Council three months in advance. Such notice shall include a statement of the extraordinary events it regards as having jeopardised its supreme interests.'

6.13 The effect of the NPT appears to me to be of prime importance in the evaluation of the non-proliferation question. Having quoted from it, I now deal with such effect. Article I clearly does not, in its first part, prevent the transfer of plutonium. Plutonium is neither a nuclear weapon nor an explosive device. It was, however, suggested that the supply of plutonium would or could amount to assisting a non-nuclear-weapon state to manufacture nuclear weapons or other explosive devices, and that it would or could, therefore, be a breach of Article I to supply plutonium to any other than a nuclear-weapon state. By parity of reasoning it would follow that a non-nuclear-weapon state would be in breach of Article II if it sought to have its spent fuel reprocessed and the plutonium returned to it, because possession of the plutonium would in fact be of assistance in the manufacture of nuclear weapons, even if the plutonium were intended for use and used entirely for peaceful purposes. Such an argument without any qualification is difficult to understand for, albeit not so directly as in the case of plutonium, the supply of uranium ore or enriched uranium also provides assistance in the manufacture of nuclear weapons. Recognising this difficulty Professor Wohlstetter suggested in evidence that the provisions of Articles I and II should be read as applying to the supply of anything which could be used for military purposes without timely warning, ie without there being time for detection and the exertion of diplomatic pressure. If the provisions were so read the embargo would not then apply to the supply of uranium or slightly enriched uranium but would apply to the supply of plutonium. That the Treaty has not been so understood is clear. Were it so read the considerable exports of plutonium both by the UK and the United States to non-weapon states, to which I have already alluded, would all have been in breach of the Treaty, as would their receipt. No-one at the time they were made apparently thought that this was the case.

6.14 The suggested construction of Articles I and II, which do not specifically refer to special fissionable material, has to be considered in the light of the provisions of Articles III and IV. Article III, by imposing on non-nuclear-weapon states the obligation to accept safeguards designed to prevent the diversion of nuclear energy from peaceful uses and applying such safeguards to source or special fissionable material, whether it is being produced, processed or used in such non-nuclear-weapon states, appears to be a clear recognition that the production and use of special fissionable material by non-nuclear-weapon states was accepted. Moreover Article III(2) specifically deals with the supply of special fissionable material to non-nuclear-weapon states and prohibits such supply except subject to the safeguards provided for by Article III(1). Such supply can therefore hardly have been intended to be within the embargo.

6.15 Article IV(1) does not appear to me to affect the argument either way. The recognition, which it contains, that all parties have an inalienable right to develop research, production and use of nuclear energy for peaceful purposes without discrimination is qualified by the words 'in conformity with Articles I and II'. If, therefore, Articles I and II are to be read as suggested, the inalienable right would also have to be read as qualified by some such words as 'provided that no such research, production or use puts a party in a position to manufacture a nuclear weapon without timely warning'. Article IV(2) does, however, throw further light on the matter and is of special importance because it contains a positive obligation with a correlative right:—

1. Each party has an obligation and a right to participate in the fullest exchange of equipment, materials and scientific and technological information for the peaceful uses of nuclear energy.

2. Each party is obliged to co-operate in contributing to the further development of the applications of nuclear energy for peaceful purposes especially in the territories of non-nuclear-weapon states.

Since the production of plutonium by reprocessing and its use in fast breeders was at the time of the Treaty the accepted future, I find it difficult to see how it can be argued that any party, whether a nuclear-weapon or non-nuclear-weapon party, has not the right (a) to develop and use reprocessing for the production of plutonium (b) to develop and use the fast breeder (c) to have access to the technology and equipment for creating reprocessing facilities and (d) to have access to reprocessing facilities which may exist in the territory of another party and to the plutonium produced by the use of such facilities. I also find it difficult to see how a party, which has developed reprocessing technology or created reprocessing facilities, would be otherwise than in breach of the agreement, if it both refused to supply the technology to another party and refused to reprocess for it.

6.16 It was submitted on behalf of FOE that the Treaty could not be construed so as to impose an obligation of this nature, at all events if it involved economic loss. This argument appears to me unsustainable. The NPT is on its face a straightforward bargain. The essence of that bargain was that, in exchange for an undertaking from non-nuclear-weapon parties to refrain from making or acquiring nuclear weapons and to submit to safeguards when provided for peaceful purposes with material which was capable of diversion, the nuclear weapon states would afford every assistance to non-nuclear-weapon states 'in the development of nuclear energy'. This, in the light of surrounding circumstances, must surely have included the development of reprocessing, the production of plutonium thereby and the use of the fast breeder. That the bargain might involve nuclear-weapon states in expense or loss is not surprising. Such expense or loss is a natural price for securing the undertaking from non-nuclear-weapon states not to become such states.

6.17 If it were necessary or indeed permissible for me to decide whether one or more parties to the Treaty could, without breach, deny reprocessing technology, reprocessing facilities or reprocessing fruits to other parties, or could, without breach, seek to coerce other parties into abandoning reprocessing and the FBR by withholding or threatening to withhold supplies of uranium or enriched uranium for their existing reactors, I should have little hestiation in deciding that it could not. In the context of proliferation risks, however, what is as or more important than the words used, clear as they appear to me to be, is the spirit of the Treaty.

6.18 About this there can, I think, be no doubt. I quote from the transcript of the evidence of Mr Patterson (FOE) when being questioned by me.

'Q. I think the last thing that I wanted to ask you was this, the Non-Proliferation Treaty came into existence at a time when everybody was looking, I think I am right in saying, to the fast breeder using plutonium as a fuel as being the long-term concept, right?—
A. Pretty generally, yes, with the usual exception of Canada.
Q. At that stage, with possibly the exception of Canada, that was seen as the long-term future?—
A. Yes.
Q. So that when parties signed that Treaty and the nuclear powers undertook to supply source materials to others for peaceful use that inevitably would appear to have contemplated providing plutonium to others for peaceful use, because that was the future which everybody then saw. It may have been foolish but would you agree that that must have been the case?—
A. I think that was certainly the intention, yes, as I understand it.
Q. Therefore it must follow must it not that a policy, by whomsoever it is operated, which denies plutonium to others is at any rate in breach, as the Japanese Foreign Minister said, of the spirit of Article 4?
—A. Certainly of the spirit of Article 4, yes'.

My reference to the Japanese Minister's statement is to a statement by Mr Sosuke Uno, Minister for Science and Technology in Japan and Chairman of the Japanese Atomic Energy Commission, made on the 31 May 1977 in which he said

'... supposing that the technology of reprocessing and plutonium use were to become the exclusive property of the nuclear weapons states, being denied to others, this would be contrary to the spirit of Article 4 of the Nuclear Non–Proliferation Treaty (NPT), which guarantees every nation an equal right to the peaceful use of nuclear energy.'

6.19 Before coming to recent events, the scope of the development of nuclear power in the world outside the Communist countries must be noticed. It can best be summarised in a passage from the evidence of Mr Patterson when being cross-examined by Lord Silsoe for BNFL.

'Q. Could I ask you, please, to turn to another country's position, Japan, and ask you to take document 238.—A. Yes, I have it.
Q. This is a speech delivered, as appears at paragraph 1, to representatives of the foreign press on the subject of atomic energy by Mr Sosuke Uno, the Minister of State for Science and Technology, and the Chairman of the Atomic Energy Commission of Japan. There are just five passages I would ask you to look at here and I would ask you to comment on.
At page 2, in paragraph 3, he says this:
'Since President Eisenhower's call for 'Atoms for Peace' in 1953 and the First International Conference on the Peaceful Uses of Atomic Energy in 1955, Japan has received from the United States the light-water reactor technology and a supply of nuclear fuels, such as enriched uranium, under the US-Japan Atomic Energy Co-operation Agreement.
"Further, with the United States' understanding, Japan has formulated its atomic energy policy on the basis of reuse of the plutonium and depleted uranium obtained by reprocessing spent fuel. To

this end, over the past two decades we have committed national appropriation of nearly three billion dollars to research and development."

Now, what the Minister is saying, and I dare ask you whether you have any disagreeing comment on it, is that his country, has, with the full knowledge and understanding of the United States, formulated its atomic energy policy on the basis that spent fuel will be reprocessed and depleted uranium and plutonium reused and that they have spent a very large sum of money to that end?—A. Quite so, I think the same is true for all countries that were encouraged into civil nuclear technology with the single exception of Canada'.

6.20 It is against this general background that one comes to current US policy and reactions to it. This policy was referred to as President Carter's policy and, in moments of enthusiasm, as President Carter's 'great initiative' or 'great moral lead'. It should however not be forgotten that the policy had its birth in President Ford's statement of October 1976:

'I have decided that the United States should no longer regard reprocessing used nuclear fuel to produce plutonium as a necessary and inevitable step in the nuclear cycle and that we should pursue reprocessing and recycling in the future only if they are found to be consistent with our international objectives.'

It was strenuously urged that this country should follow that policy because failure to do so would increase proliferation risks. To follow the policy would, it was said, involve refusal of planning permission for THORP.

6.21 The policy was developed by President Carter at a news conference on 7 April 1977. It comprised in essence the following:—

1. Indefinite deferment of commercial reprocessing and recycling of plutonium.
2. Giving increased priority to the search for alternative designs for the FBR and deferring the date when FBRs would be put into use.
3. Increasing US capacity to provide adequate and timely supplies of nuclear fuels to countries that needed them '*so that they will not be required or encouraged to reprocess their own materials.*'
4. Proposing to Congress the necessary legislation to sign supply contracts and *remove the pressure for the reprocessing of nuclear fuels by other countries which did not then have that capability.*
5. An embargo on the export of equipment or technology that could permit uranium enrichment or chemical reprocessing.
6. Pursuing discussions of a wide range of international approaches and frameworks that would permit all countries to achieve their own energy needs while at the same time reducing the spread of the capabilities for nuclear explosive development.*

Under the last heading the President mentioned the establishment of an International Fuel Cycle Evaluation Programme (INFCEP) 'so that we can share with

*The italics are mine.

countries which have to reprocess nuclear fuel the responsibility for curtailing the ability for the development of explosives.' The INFCEP has since then been established. The President also mentioned that the US would have to help to provide some means for the storage of spent fuel and, since that time, plans have been announced for the US to receive and store such fuel.

6.22 Certain remarks made by the President at this news conference are of significance in the context of the question whether permission for THORP, and its building pursuant to such permission, would run counter to US policy. I quote them:—

a. 'We are not trying to impose our will on those nations like Japan, France, Britain and Germany which already have reprocessing plants in operation.'
b. 'Obviously, the smaller nations, the ones that now have established atomic power plants, have to have some place either to store their spent fuel or to have it reprocessed and I think we could very likely see a continuation of reprocessing capabilities within those nations that I have named and perhaps others.

   We in our own country do not have this requirement. It is an option that we might have to explore many, many years into the future.'
c. 'I hope that by this unilateral action we can set a standard and that those countries which don't now have reprocessing capability will not acquire that capability in the future.'
d. 'The one difference that has been very sensitive, it relates to, say, Germany, Japan and others is that they feel that our unilateral action in renouncing the reprocessing of spent fuels to produce plutonium might imply that we criticise them severely because of their own need for reprocessing. This is not the case. They have a perfect right to go ahead and continue with their own reprocessing efforts. But we hope that they will join with us in eliminating in the future additional countries which might have had this capability evolve.'

6.23 It is clear that, when the President was acknowledging the right of countries such as ours to continue reprocessing, he referred to reprocessing for home use of the plutonium only. It would be absurd to object to the export of reprocessing capability to nations which do not have it, but to have no objection to the export of plutonium itself. Nevertheless it appears to be clear that the building of THORP itself would not be counter to US policy so long as no plutonium produced by it was exported. So limited there would be no direct increase in proliferation risks.

6.24 If the use of THORP were not so limited and plutonium were supplied to non-nuclear-weapons states it would not be so supplied until, at the earliest, 10 years from now, for THORP would not be operative until then. The effective risk would thus be a risk of increased proliferation, at the earliest, in 10 years time. In the meantime the incentive to customers to develop their own facilities would be reduced by the knowledge that they could send their spent fuel here, have it reprocessed and have the plutonium required for fast breeder programmes returned to them, either as plutonium or in the form of fuel rods.

On the other hand suppose that the use of THORP is limited, and that nations with the capability to reprocess deny it to others, the incentive to others to develop their own capability must immediately be increased. US policy clearly acknowledges this by its inclusion of the need both to assure supplies of enriched uranium and to provide storage for spent fuel. The question which therefore arises is whether these two provisions would be effective to nullify the increased incentive which denial by itself would produce.

6.25 The civil incentive to reprocess is the achievement of resource independence, for a country which depends for its nuclear reactor fuel supplies on imports, is in a vulnerable position both financially and politically. The disadvantage of becoming too dependent on importing oil supplies has been all too effectively demonstrated in recent years and it was submitted to me that, under present circumstances, countries with no reprocessing capabilities could be forced to stop the development of such capability, if the countries upon whom they relied for uranium supplies or enrichment services joined in withholding supplies from them. Such a sanction is undoubtedly a powerful one. It could also be used to enforce the acceptance of policies other than non-proliferation. Limitation of reprocessing would prevent the resource independence which is legitimately sought by nations without their own supplies. Furthermore if, at the same time as foregoing reprocessing, such nations were to send their spent fuel to the United States (or to other nations with an existing capability) for storage, they would be depriving themselves of an existing capability to become resource independent. If the spent fuel is retained the possibility of so becoming remains.

6.26 It must be at least doubtful if assurances of enriched uranium supplies and the acceptance of spent fuel would or will relieve the pressure, particularly when withholding of reprocessing technology and services is, at the least, against the spirit of the existing NPT, and would render abortive the very large expenditures encouraged by the initiator of the policy of denial. What guarantee could there be that the assurance of enriched uranium supplies would not itself be ignored at some time in the future? Might not America and the other nuclear-weapon states have yet another change of policy and ignore undertakings to provide enriched uranium? Other countries might ask themselves such questions as these. If they did, the response to the policy might well be a marked acceleration in the development of reprocessing capability as an insurance against future changes in policy. If this were to happen, then, before ever THORP could have produced a single kilogram of plutonium, several other countries might well have produced their own.

6.27 I have already mentioned the Japanese reaction to the policy. I should also mention that the Commission of the European Communities in its communication to the Council of the EEC on 2 July 1977, entitled 'Points for Community Strategy on the Reprocessing of Irradiated Nuclear Fuels' (G30) advocated the development of reprocessing and considered it to be compatible with non-proliferation.

6.28 It must also be remembered that it may be necessary in some cases to reprocess spent fuel and this is recognised by current US policy. On 1 July 1977 the Deputy Assistant Secretary of the US Bureau of Oceans and International and Scientific affairs wrote, in a letter to the Attorney for the NRDC:

'In reponse to your inquiry of US policy governing requests we receive to approve the retransfer of US-origin spent fuel for reprocessing, our policy is that each such request will be considered on a case-by-case basis, with approval contingent on a clear showing of need, such as spent fuel storage capacity problems.'

One such permission dated 16 September 1977 has already been granted to Japan to transfer 8.3 tonnes of spent fuel to BNFL for reprocessing on the basis that this was vitally necessary to maintain a particular power station in operation, the spent fuel storage capacity being full. It is interesting to note that one of the conditions attached was:

'That this spent fuel is to be retained by BNFL until... reprocessing and that thereafter the produced plutonium will be returned to Japan... In accordance with applicable agreements for co-operation such transfers would, at that time have to be approved by the Government of the United States.'

Professor Wohlstetter had accepted, prior to the issue of this permission, that permissions would be given in some cases, that there must be a reliable plant somewhere, and that both France and the United Kingdom were possibly suitable locations. This was also accepted by Professor Rotblat.

6.29 How many permissions there will be and for how much spent fuel it is impossible to assess. It appears that they will be given where storage capacity has expired, at least until additional storage capacity has been created somewhere. They may also have to be given in cases where the condition of fuel on leaving a reactor, or after a period of pond storage, is such that storage or further storage is undesirable. With so much spent fuel arising in the course of the next two decades it is clear that there should be adequate and reliable reprocessing facilities with spare capacity somewhere and that the obvious locations for such facilities are in one or more of the present nuclear weapon states.

6.30 If such facilities are created their creation will not, as I see it, increase the proliferation risk unless either
   a. their creation necessarily involves, or is treated by others as necessarily involving, a commitment to plutonium-using FBRs or
   b. the plutonium produced by the facilities is returned to fuel owners in a form which will enable the owner country to proceed to a bomb without time for diplomatic pressure to be exerted.

6.31 In view of the amount of plutonium which will in any event be produced from magnox reprocessing and the plain need to have a reprocessing plant somewhere, the creation of THORP could not in my view reasonably be seen as a commitment of the kind mentioned. Indeed it is hard to see how any such commitment could be made until a commercial FBR had been built and successfully operated for some years. When that stage had been reached, but not until then, would a country know whether it could, even if it wanted to, commit itself to a plutonium-using FBR programme.

6.32 Returning the plutonium to non-nuclear-weapon owner countries will represent an increased risk, but this might be mitigated by returning only when required for civil reactors and then only in the form of briefly irradiated fuel rods.

6.33 Whether this risk, which will not arise for at least 10 years, is or is not a greater risk than the increased incentive which the denial of technology and facilities would immediately create, is a matter which I cannot assess. Its assessment is a matter for the Government and depends amongst other things on information on the reactions of other countries to the policy. The argument that the grant of permission would add to proliferation risks was not however established before me. Indeed I would go further. Since (i) there will be no direct risk arising from THORP for at least 10 years (ii) to deny reprocessing facilities would be against the spirit – and as I think the letter – of our obligations under the main existing bulwark against proliferation (iii) the denial of such facilities would create an immediate incentive to others to develop their own facilities (iv) there is a world need for adequate reprocessing facilities somewhere, it appears to me that a grant of permission would have a non-proliferating effect rather than the reverse. I do not accept that the best way to achieve a new bargain is to break an existing one.

6.34 It may be that, if permission is granted, INFCEP will thereafter result in an international agreement not to reprocess commercially. If it does, it does not follow that THORP would then be redundant. The accumulation of ever larger stocks of spent fuel in the world, without facilities available to reprocess considerable quantities should some unforeseen problem render it so necessary, would in my view be, at best, imprudent, and, at worst, irresponsible. Even, however, if THORP did become redundant, and I do not consider that it would, this would merely mean that some expenditure would have been wasted. This is an event which may always happen when plans are made to cover contingencies many years ahead. The expenditure may be regarded as an insurance premium.

7. Terrorism and Civil Liberties

7.1 This is a matter upon which the evidence which could be tendered before me was very limited. Rule 10(4) of the Procedure Rules provides:

'The appointed person shall not require or permit the giving or production of any evidence, whether written or oral, which would be contrary to the public interest.'

It appeared to me clear, and I so announced at the preliminary meeting, that I could not therefore allow any evidence which would prejudice national security either by disclosing our own defence measures or by providing information which might assist others to develop a nuclear weapons capability: nor could I allow any evidence which might assist a terrorist organisation to gain access to – or claim to have gained access to – special nuclear materials. At the meeting I said:

'Everyone will, I am sure, readily understand this, particularly perhaps those whose objections are most strongly felt. It would, for example, be neither in their interests nor the public interest, nor indeed the interest of mankind, if, in the process of endeavouring to secure or securing a rejection of BNFL's present application, they were to secure the disclosure of information which would, or might, create or increase the nuclear capability of others, or enable others over whom we have no control to create their own supplies of plutonium, or, finally, which would or might expose existing or possible future installations in this country to vulnerability from terrorists.'

7.2 All parties, as I had expected, fully accepted the position so far as the Inquiry was concerned but it necessarily precluded me from investigating existing physical security precautions, both at the plant itself and during transport of plutonium to other locations in the United Kingdom and elsewhere, and the extent of existing precautions for vetting personnel working at the plant or involved in transport, their families and friends. It precluded also any investigation of the precautions which would be taken under these heads should outline planning permission be given and acted upon.

7.3 Grave anxieties were expressed by objectors under both heads, the starting point being, as it was in many instances, the Sixth Report. Both matters are dealt with in Chapter VII of that report and in paragraph 532 of Chapter XI – the Summary of Principal Conclusions and Recommendations. The principal matters with which the Royal Commission was concerned were:

1. Release by sabotage of the contents of the HAWs (para 309).
2. The release by sabotage of the contents of a spent fuel flask (para 313).
3. The theft of plutonium and its use or threatened use in a bomb (para 325).
4. The extent of the secret surveillance of members of the public and workers which might be necessary (para 332).

It is convenient to quote paragraph 532 in full:

'532. Security and the safeguarding of plutonium

22. Plutonium appears to offer unique potential for threat and blackmail against society because of its great radiotoxicity and its fissile properties (182, 322, 323).
23. The construction of a crude nuclear weapon by an illicit group is credible. We are not convinced that the Government has fully appreciated the implications of this possibility (325).
24. Given existing or planned security measures, the risks from illicit activities at the present level of nuclear development are small; the concern is with the future (308).
25. Plutonium extracted from fuel reprocessed for a foreign customer should, if returned, be incorporated in new fuel elements (319).
26. The unquantifiable effects of the security measures that might become necessary in a plutonium economy should be a major consideration in decisions on substantial nuclear development (332, 335). Security issues require wide public debate (336).'

7.4 Both matters are dealt with in the White Paper. In summary what was then said by way of response was

1. 'The Government agree that decisions on the form in which plutonium is to be returned must be based on non-proliferation and security considerations. They will be pursuing the question in further international discussions on non-proliferation, including the proposed international fuel cycle evaluation (para 33).'
2. With regard to the risk of construction of a crude illicit weapon, 'security measures in connection with the transport and storage of plutonium have been greatly strengthened over the last two years and will be reviewed at regular intervals.'
3. Interference with civil liberties could be reduced by designing security into nuclear plants and the Government would ensure that full attention was

given to this at the planning and design stage.
4. The degree of surveillance needed depends more on the prevalence of terrorism than on the availability of plutonium but the Government would continue to preserve civil liberties in the nuclear field as in other fields.

7.5 This response does not materially advance matters in the sense that it does not give further information to the public. To a large extent the matter is one in which the giving of further information would or could tend to increase the risks. I take, by way of example only, the question of the possibility of constructing a nuclear device if plutonium were obtained. The Royal Commission found it credible. It was advanced on behalf of Windscale Appeal (WA) by way of the evidence of Dr C M H Pedler, MD, BS, PhD, MC Path that it would be not only credible but easy. WA co-operated with my suggestion that this evidence should not be read but simply handed in. This course appeared to me desirable for, if the claim was unsustainable, it could only be shown to be so by a line of cross-examination which would or might reveal how to surmount a difficulty or difficulties which Dr Pedler had not appreciated. I make no finding as to the ease with which a crude nuclear device could be made by terrorists. I accept the Royal Commission's view that it is credible.

7.6 On this aspect of the application I draw attention to the following:
  a. Although plutonium has been produced and moved both intra- and inter- nationally for over 25 years there has not been any terrorist abstraction or threat so far as is known.
  b. Dr Cochran stated in his proof of evidence that there were technical fixes 'that could all but eliminate the threat of plutonium theft and subsequent construction of nuclear bombs by terrorists and the like'. I also quote from his evidence:
     'Q. Supposing that the plutonium arising from reprocessing is returned to the country which submits the spent fuel in the form of mixed oxide fuel for a thermal reactor which has already been to some degree irradiated. That I think you agreed, would be a form of technical fix which would be effective against the non-state adversary, is that right.
     A. This is correct.'
  c. Dr Cochran was of the view that improved measures to safeguard plutonium could be incorporated in a modern up-to-date plant.
  d. There was no evidence that at present the safeguarding of plutonium has constituted any undue interference with civil liberties.

7.7 The technical fixes discussed varied from 'spiking' the container, i.e. including in it some highly radioactive substance which would make it dangerous or possibly lethal for an unauthorised person to open the container without special shielding and handling facilities, to the supply of the plutonium as mixed oxide fresh fuel rods but after brief irradiation to ensure that the rod assemblies were highly radioactive. It was this last method which Dr Cochran regarded as being effective against terrorists. I am satisfied that it would be the most effective of the possible 'fixes'. I am also satisfied that it would, if it could be introduced, virtually eliminate the terrorist risk. Its introduction would present certain difficulties both to BNFL and to the receiving customers but I agree with the Royal Commission's recommendation and I am satisfied that the difficulties could be overcome albeit that to do so would be costly. There is no provision in the proposed Japanese contract which would entitle BNFL to insist on return in this form but the Government could impose such a requirement should it still appear desirable when the situation arises.

7.8 The Royal Commission, when considering terrorism and civil liberties, as with many other matters, were concerned principally with the implications of what they call a 'plutonium economy' perhaps 50 years in the future. This appears clearly in three paragraphs which I quote:—
  (a) 'We are satisfied that in this country at least, with the present state of nuclear development and with the security measures that are now in force or are being introduced, the risks to society from illicit activities are small. The main concern lies with the future in which there could be a substantial growth in nuclear power and a move into the "plutonium economy" '. (para 308).
  (b) 'There would be a particularly serious threat (of economic damage) if in the future we came to depend upon plutonium as our main source of energy'. (para 311).
  (c) 'We find it hard to believe that such an intolerable situation (widespread surveillance) could arise in this country, though it might do so in countries with oppressive regimes. It must be remembered, however, that in considering the hazards of the plutonium economy we are concerned with conditions as they might be fifty or more years ahead. What is most to be feared is an insidious growth in surveillance in response to a growing threat, as the amount of plutonium in existence, and familiarity with its properties, increases; and the possibility that a single serious incident in the future might bring a realisation of the need to increase security measures and surveillance to a degree that would be regarded as wholly unacceptable, but which could not then be avoided because of the extent of our dependence on plutonium for energy supplies. The unquantifiable effects of the security measures that might become necessary in the plutonium economy of the future should be a major consideration in any decision concerning a substantial increase in the nuclear power programme.' (para 332).

7.9 I would not regard any long term assessment as forming any part of the present Inquiry unless it were shown that permission for and the building of THORP necessarily constituted a commitment to a large scale reliance on nuclear energy produced by reactors using plutonium as a fuel. I avoid the use of the phrase 'plutonium economy' for I consider it to be both emotive and inaccurate. It was not shown to me that permission, if granted, would involve such a commitment, although it was submitted that this was so. It was also submitted also even if permission would not itself involve such a commitment it would render such a commitment the more likely. To an extent this is true. The building and operation of THORP would make more plutonium and more uranium available for use as fuel and would provide more experience in reprocessing. It might therefore be possible to make a larger commitment when the time arises than it would be if THORP were not built. If a larger commitment is possible, the likelihood of making the larger commitment must to some extent increase, but the matter goes no further than this and forms, in my view, no justification for trying to make an assessment now of the effect upon civil liberties of a commitment which could not be made for at least 15 years and might never be made at all (see para 7.19 below).

7.10 If then the matter for consideration is the effect of THORP itself, what does it involve from a security and civil liberties point of view? It involves the following:
(a) There would be separated at Windscale about 40 tonnes of plutonium in addition to the 50 or more tonnes which would arise from magnox reprocessing.
(b) About half of that additional amount would be stored at Windscale, for future use in either thermal reactors or FBRs. This fuel would be fabricated at Windscale.
(c) The other half would be stored for a time and then despatched either as plutonium or as mixed oxide in fuel rods for thermal reactors or FBRs, possibly briefly irradiated.
(d) There would be more highly active waste in store in HAWs.

7.11 As to (a) and (b) above, I do not see any significantly increased risk and this was largely accepted by Dr Cochran. As to (c) above the risk could be largely eliminated by irradiation or other technical fixes (see para. 7.6). As to (d) above, if the highly active waste is not created there will be increased storage of spent fuel which carries its own dangers and which contains all the plutonium instead of a small fraction of it. I do not see any significant increase in risk in this connection. There remains the question of civil liberties. It was not seriously suggested that THORP alone would involve any significant interference with civil liberties and I do not consider that it would.

7.12 It follows from the above that I do not, on the evidence which I was able to take, see any reason to suggest that outline permission should be refused either on security grounds or on civil liberties grounds. I stress however that I was not able to take any detailed evidence touching either matter.

7.13 Arising out of this constraint, it was submitted by or on behalf of a number of objectors that there should be a separate Inquiry into the security measures which would be necessary, in particular by way of surveillance, if there was a large commitment to plutonium fuelled reactors, not merely to see if such measures would be adequate to protect against terrorism but also, and especially, to determine what inroads on civil liberties they would make, and whether such inroads would be tolerable. This was based mainly on the argument that the growth of interference was slow and insidious and that failure to examine the matter fully before the first step was taken would or might mean that, before the full effects had become apparent, the country would be so dependent on plutonium fuelled reactors that it would be too late to turn back. Such an Inquiry, it was submitted, should take place before any decision to grant planning permission for THORP was made. The suggestion that the implications of commitment to a large scale plutonium fuelled programme should be considered arose firstly because it was very difficult to suggest that THORP by itself would be likely to involve any serious increase in interference with civil liberties. It arose in part also because of the suggestion that a commitment to THORP necessarily involved the larger commitment as well.

7.14 It was also submitted that it was essential that the security structure should be such that there should be no doubt about the person ultimately responsible and that such person should be directly answerable to Parliament. To this end it was suggested that two crown servants should be on the Board of BNFL, so that there would be direct Ministerial responsibility. These matters should also, it was contended, be investigated in the suggested Inquiry.

7.15 I have not the least doubt that the adequacy of precautions planned to be taken, and in fact taken, to protect against terrorism and the effect of such precautions on civil liberties are matters of concern. They clearly are and the Government has recognised them to be so. I do not however accept the argument that there should be an Inquiry into them before a decision on the present application is made.

There appear to me to be two quite separate areas of investigation namely:
(1) The adequacy of measures to be taken so far as THORP itself is concerned and their effect on civil liberties.
(2) Measures and their effects on civil liberties which might result from a large scale commitment to plutonium fuelled reactors.

As to the first, the initial step is to design into the plant as much inbuilt security as possible and thereby minimise additional precautions which might affect employees or the public. The Government has given an assurance that this will be done. I do not see any means whereby there could be a public debate into such matters for such a debate would depend on detailed evidence the disclosure of which would or might impair their effectiveness. Moreover, until the design of the proposed plant has proceeded a great deal further, it will not be known what inbuilt precautions can be or will be taken. This was broadly accepted but used as an argument that permission should not be given until the design had proceeded much further.

7.16 I do not accept this. The research and development required to reach final design is estimated to take two years and to involve the expenditure of £17 million. It appears to me mere commercial prudence to seek outline permission before embarking on such expenditure and thus to be sure that there will be no impediment on planning grounds if, when final design stage is reached, such design is acceptable not merely from a security point of view but also so far as safety aspects are concerned, a matter which I revert to later.

7.17 I consider that the argument for a preliminary Inquiry covering the first of the areas mentioned is unsustainable. It does appear to me, however, that there is much to be said for a system involving initial independent checking of security precautions taken and to be taken and their subsequent review from time to time. There is at present machinery for independent investigation of breaches of security which have occurred. The Standing Security Commission was set up in 1964 with the following terms of reference.

'If so requested by the Prime Minister, to investigate and report upon the circumstances in which a breach of security is known to have occurred in the public service, and upon any related failure of departmental security arrangements or neglect of duty; and, in the light of any such investigation, to advise whether any change in security arrangements is necessary or desirable.'

It has seven members and its Chairman is Lord Diplock. As can be seen from the terms of reference, not only is it limited to dealing with breaches of security which have occurred but it is also limited to investigation of breaches of security within the public service. Thus, even if there were a breach of security at Windscale, it would not necessarily be possible to activate it.

7.18 Of the very nature of things those who design and operate security measures will consider them adequate and effective but they may be mistaken and they may have overlooked something which an independent person or body would identify. In view of the anxiety and present level of terrorism I would recommend that consideration be given to charging some independent person or body with the task of (a) vetting security precautions, both at Windscale and during transit of plutonium from Windscale and (b) reviewing the adequacy of such precautions from time to time. This would I think do much to reassure the public and to ensure that the stable door was closed before rather than after the horse had gone.

7.19 With regard to security measures which might be necessary in the event of a large scale commitment to plutonium fuelled reactors, the possibility of such a commitment does not appear to me to arise unless and until

(1) It is decided to build the first commercial FBR (CFR 1) and
(2) CFR 1 has been built and successfully operated.

Only at that stage would it be known whether the country can, if it wants to, embark on a large scale FBR programme with the advantage of resource independence that such a course would involve. By that time – some 15 years ahead – much will have happened. It might be that by then it had become apparent that any further use of nuclear power by FBRs or thermal reactors was unnecessary, at least for a further decade or two. Alternatively it might then be clear that, short of a large and immediate commitment to FBRs, with whatever erosions of civil liberties might go with it, the country would have to accept a severe reduction in living standards or a greatly increased pollution of the environment from coal fired stations and a far greater dependence on a particular fuel source and those whose co-operation is required for its exploitation than presently exists. Only at that time (if it ever comes) and in the light of the circumstances then prevailing would the sort of Inquiry envisaged appear to me to be of any real value; for what would be acceptable in the field of erosion of civil liberties must depend on what the alternatives are, or are estimated to be, at the time when the question arises. All that one can presently say is that, in 15 years time, known and estimated alternatives will be different from what they now are or appear to be.

7.20 Finally in this section I deal with a matter which appeared to cause much concern to objectors. It arises in this way:

(1) In February 1977 a number of questions were submitted to the Secretary of State for Energy on behalf of FOE, the Council for the Protection of Rural England (CPRE) and the National Council for Civil Liberties (NCCL). These questions included:

'(6) Is it anticipated that bodies opposed to the development of nuclear power will be subject to security surveillance? If so what criteria would be used to determine which bodies should be so subject?'

(2) The answer given was:

'Bodies and individuals opposed to the development of nuclear power would not be subject to surveillance unless there was reason to believe that their activities were subversive, violent or otherwise unlawful.'

(3) In evidence Mr Herzig when asked to define the word 'subversive' quoted from an answer given in the House of Lords:–'subversion is defined as activities threatening the safety or wellbeing of the

State and intended to undermine or overthrow Parliamentary democracy by political or violent means'.

7.21 This definition was severely criticised as being so vague and so potentially wide in its possible interpretation that it left almost limitless scope for administrative discretion, a discretion for the exercise of which there was no effective Ministerial responsibility. This, it was argued, was particularly serious since there is no remedy in law for an interference with civil liberties as such. The word 'subversion' should not therefore be used, or if used should be much more narrowly defined. The argument was chiefly developed on behalf of Justice and NCCL.

7.22 The underlying anxiety expressed is readily understandable, the more so since the nature of security is such that an innocent person who is harmed, e.g. by being rejected for some employment, may never know the reason for such rejection and can therefore never have the opportunity to disprove what has been considered to render him unacceptable. Even if he does discover and can disprove the supposed fact, he will or may have no remedy.

7.23 Whilst I accept fully that there is cause for anxiety I am unable to see that it will be dispelled, either by abolishing the use of the word 'subversive' or by defining it more narrowly. We live in a world in which there are people with evil purposes and more specifically in a country whose system of government and whose operation of the rule of law has taken centuries to build. There are beyond doubt some who would wish to see that system and operation swept away. When such people move towards their ends by violent or other unlawful means no-one suggests that they should not be checked. But the progress of such people towards their objective can be as slow and insidious as the invasion of civil liberties which may be necessary to defeat their ends. There is little to be gained by the preservation of civil liberties if the doing of it results not only in the insidious destruction of a system of government and law which values such liberties, but also in its replacement by a system which will then destroy the very civil liberties which it is desired to preserve. It is, of course, equally the case that it would be profitless to erode civil liberties to a point where they were as restricted as they would become in the event of the success of those whose aims the erosion was designed to defeat.

7.24 The problem is easy to state but there is no easy solution. Indeed I can see no solution at all. If the sort of activities under consideration are to be checked, innocent people are certain to be subjected to surveillance, if only to find out whether they are innocent or not. Equally certainly friends and relatives will be subjected to distasteful and embarrassing enquiries. The most that one can do, as it seems to me, is to require that the Government should ensure that the interference with our liberties goes no further than our protection demands and that there should be some Minister answerable to Parliament if interference goes further than this.

7.25 The stating of the requirement, like the stating of the problem, is relatively easy, but the fulfilment of the requirement, like the solution to the problem, presents great difficulties. I do not for a moment suggest that any of the objectors before me were motivated by a desire to harm this country, but it is plainly possible that the aim of doing harm can be pursued under the outward guise of furthering such a worthy aim as the protection of civil liberties. A campaign to lessen surveillance, ostensibly to preserve civil liberties, could therefore be mounted by people whose aim was not the preservation of such liberties but increased opportunity to further their own destructive ends. I mention this because one party advanced the view that much of the opposition to nuclear power in general and to THORP in particular had as its objective, not the preservation of the health and wellbeing of present and future generations and of the natural environment, but economic and other damage to this country. Whilst I accept the possibility that opposition could have such an objective and that it is a possibility which must be kept in mind, I reject the submission that any of the parties before me were so motivated. Some, it is true, were in favour of a reduced standard of living, if this were necessary to avoid reliance on nuclear power, but this was freely acknowledged – indeed it was specifically advanced as desirable. I mention in passing that there was no evidence before me which went even a small way towards establishing that the country at large would be prepared to accept such an alternative.

8. The Need for Reprocessing of Oxide fuel and Relationship to the FBR Question

8.1 It is convenient to deal first in this section with the second matter raised in the heading, for the facts relating to it are clear. The present stocks of plutonium from magnox fuel reprocessing, together with the additional plutonium which will arise from reprocessing such fuel until the end of the useful lives of the magnox stations, will total some 55 tonnes. Should it be decided to build CFR 1 to test the capabilities of the type it is likely that it will have a generating capacity of 1.25 Gigawatts (Electrical) (GW(E)). On current estimates the initial charge for such a reactor would require some 3.4 tonnes of plutonium from magnox fuel. This initial charge could be met from existing stocks of plutonium which amount to $7\frac{1}{2}$ tonnes. How much more would be required to maintain it in operation would depend upon the stage at which the reprocessing of spent fuel discharged from it was introduced. But even if reprocessing were not begun until 10 years after start up the requirement would be no more than about 22 tonnes. Hence it is clear beyond doubt that oxide fuel reprocessing is not required in order to preserve the option to build CFR1. It is also clear that plutonium from magnox reprocessing would enable a start to be made on an FBR programme, following upon CFR1 should it be decided to embark upon such a programme.

8.2 The size of the FBR programme which would be feasible without the need for any contribution from oxide fuel reprocessing depends upon the assumptions which are made as to reactor size, the time at which the reprocessing of FBR fuel is introduced and the speed at which spent fuel emerging from the reactors can be reprocessed and the recovered plutonium fed back to the reactor. BNFL suggested a possible programme based on the assumptions that each FBR would have a generating capacity of 1.25 GW(E), that reprocessing of FBR fuel could be introduced at the outset and that spent fuel could be reprocessed and the recovered plutonium fed back into the reactors within a year. On these assumptions they estimated that the total requirement of magnox plutonium would be 5 tonnes per GW(E) of reactor capacity or 6.25 tonnes per reactor of the size assumed.
    The programme suggested as possible was:

CFR1	1.25 GW(E)	on line 1990
2nd Reactor	1.25 GW(E)	on line 1994
3rd Reactor	1.25 GW(E)	on line 1998
4th Reactor	1.25 GW(E)	on line 1999
5th & 6th Reactors	2.5 GW(E)	on line 2000
7th & 8th Reactors	2.5 GW(E)	on line 2001

Thereafter 2 Reactors per year

8.3 BNFL accepted that magnox plutonium could support such a programme up to and including the 8th Reactor but no further. It will be observed that the suggested programme envisages the 2nd reactor coming on line four years after CFR1. This, bearing in mind building times, must imply considerable building work having been done upon it before CFR1 had been started up, let alone tested. This appears to me unlikely. The interval between CFR1 and the 2nd reactor coming on line would, I consider, be considerably longer than suggested. In other respects, the programme might be shortened. In the present context, however, the time span is not important. The number of FBRs which could be fuelled from magnox plutonium would remain the same, unless the suggested programme were so much extended in point of time that a substantial increment of plutonium arose from the early FBRs themselves.

8.4 It will also be observed that, on the assumptions made, there would be, after the 8th reactor had been fuelled, a balance of some 5 tonnes of magnox plutonium remaining as a contribution to the requirements of the 9th reactor. By that time the earlier reactors in the programme would also be contributing plutonium but, even if they had produced enough to make up the balance required for one further reactor, there is no doubt that to continue such a programme beyond the 9th reactor would require plutonium from oxide fuel reprocessing. Plutonium produced by the existing FBRs would be insufficient for the purpose. It is estimated that the time taken for an FBR to produce enough plutonium to fuel an additional FBR would be 25 years. Even if (which is not the case) 1/25th of the amount required to fuel another reactor were produced every year it would therefore require 50 reactor years operation per year to provide enough plutonium to fuel two further reactors per year. Since at this stage there would only be, at most, 9 reactors in operation, there would only be 9 reactor years of FBR plutonium production per year.

8.5 I conclude from the foregoing (i) that it is unnecessary to embark on oxide reprocessing in order to keep open the option, not merely of CFR1 but also of a follow-on FBR programme up to 8 FBRs with a total generating capacity of 10 GW(E); (ii) that to go further than this, oxide reprocessing would be necessary at some stage, but (iii) that it would not be necessary until after 1987 which is the date presently estimated for the commencement of operation of THORP if permission is given.

8.6 For how long a start on oxide reprocessing could be delayed is a matter of conflict. BNFL accept that the programme which they envisage could be sustained if the start were delayed for five years but for no longer. FOE made the general submission, not particularly allied to the suggested programme, that a delay of at least 10 years would not prejudice any option. The possible length of any delay and the effects of delay are questions which raise a number of problems. Suppose, for example, that it should ultimately be decided that an FBR programme as envisaged by BNFL should be maintained, it would need, it is presently estimated, 12.5 tonnes of plutonium per year to sustain it. Some would be contributed by existing FBRs but the bulk would have to come from AGR reprocessing. If the requirement from this source were, say, 10 tonnes p.a., AGR fuel would then have to be reprocessed at the rate of some 2500 tonnes p.a. for the plutonium yield from AGR fuel is only about 0.4 per cent. If, therefore, there were a delay until, for the maintenance of the suggested programme, oxide fuel had to be reprocessed, BNFL would then be faced with reprocessing at four times the rate presently proposed in a plant or plants with more than double the capacity presently proposed. If a start is made as proposed, however, the stocks can be built up at a relatively low rate of reprocessing while experience is gained. I draw attention, also, to the fact that, if the annual requirement were to be 12.5 tonnes plutonium and thus entailed the reprocessing of 2500 tons per annum, the suggested programme could only be maintained, once the backlog of spent fuel had been dealt with, if there were many more AGRs to provide the spent fuel required than are now in existence, under construction or likely to be ordered in the near future.

8.7 From the above it is clear that the longer the delay is before a start is made the greater will be the rate at which reprocessing will have to take place if and when it is ultimately required and the larger will be the plant capacity required.

8.8 If oxide fuel reprocessing does not take place for a period, the spent fuel will have to be stored until either it is finally disposed of or it is reprocessed. If a start on THORP is delayed for 10 years this will mean that some 20 years will elapse before reprocessing commences, i.e. that spent fuel coming out of the reactors now will have to remain in pond storage – or some alternative – until about 1998 by which time, subject to such relief as may be given by the use of B204/5, there will be some 4000 tonnes of such spent fuel in ponds or alternative storage.* This raises the question whether pond storage for periods of 20 years or more is acceptable.

8.9 At an early stage of the Inquiry it was suggested that extended storage of spent fuel would have considerable advantages. The plutonium and uranium contained in it would be available for recovery if and when required. In the meantime there would be no separation of plutonium and thus no addition to stocks

*i.e. from reactors presently existing or under construction

or additional movements. Discharges of radioactivity to the environment which are involved in reprocessing would be avoided, certainly for a period and possibly for ever. Since the plutonium will not be needed for a considerable time in any event and since there is at present no shortage of uranium supplies, storage for long periods would, it was argued, be advantageous even if it was ultimately decided to reprocess. During the period of delay, knowledge of the potential harm from discharges would increase; methods for reducing or completely containing discharges might be developed; there would be time to increase and strengthen the safeguards against the possible diversion of plutonium to weapons use before it was supplied to non-nuclear states; more experience could be gained by the use of B204/205; alternative sources of power might, when coupled with conservation measures, satisfy energy needs to an extent which would make it unnecessary to resort to nuclear power at all. Delay would not prejudice an ultimate decision to reprocess whereas, if reprocessing began, a practically irreversible situation would be created. The highly active waste could not be restored to its original condition. Plutonium stocks would be there and would have to be disposed of and the released radioactivity could not be withdrawn from the environment.

8.10 The attractions of the foregoing argument are readily apparent. It was countered by BNFL on the basis:
1. that storage for the periods envisaged was not acceptable at any rate for AGR fuel;
2. that fuel would in the end have to be reprocessed before ultimate disposal even if this did not involve separation of the plutonium;
3. that separation of the plutonium was desirable because it could be used in thermal reactors;
4. that this would be a safer course than committing it to ultimate disposal whether in the form of spent fuel or as part of solidified waste;
5. that total plutonium stocks would thereby be reduced;
6. that delay would make the introduction of reprocessing if ultimately decided upon more difficult. There would be a lack of continuity and the dispersal of expertise and the plant ultimately required would be larger;
7. that there was a world need for reliable reprocessing capacity in one or more existing nuclear-weapon states to deal with the situation which would be created if spent fuel required to be reprocessed because of failures;
8. that delay would involve the development of new systems for storage which would be unnecessary if reprocessing was, in the end, undertaken.

8.11 Crucial to the argument for extended storage is the question whether storage for extended periods is or would be satisfactory. The argument that it would was originally based on an unqualified statement in the Sixth Report (para 377) that 'fuel clad in stainless steel or zircaloy could be stored for a few decades in ponds', and the fact that both in Canada and the USA long-term storage was apparently regarded as feasible. The statement in the Sixth Report appeared to have been based on evidence given by the United Kingdom Atomic Energy Authority (UKAEA) but enquiries revealed that this was not the case. The UKAEA had not given and were not in a position to give any such evidence.

8.12 With regard to stainless steel clad fuel the Central Electricity Generating Board (CEGB) wrote on the 25 November 1976 (CEGB 9) to four American companies in the following terms:

'*Storage of Stainless Steel Clad Fuel*
The UK Advanced Gas-Cooled Reactors make use of stainless steel clad, uranium oxide fuel, and we are seeking to identify any problems which may arise in the prolonged storage of irradiated AGR fuel under water.

I understand that you have experience of storing stainless steel clad fuel after discharge from the reactor. I should be grateful for any available information on problems which have been encountered or are anticipated, particularly where contamination could occur or where subsequent reprocessing of the fuel is difficult.'

8.13 Replies from three of the companies were put in evidence and CEGB had also pursued their inquiries by telephone. The net result was that none of the four companies had stored any such fuel for more than seven years, that they had in that time not identified any problems and that they did not anticipate any problems. Canada does not use stainless steel clad fuel.

8.14 As to zircaloy fuel, storage for longer periods up to about 15 years has taken place without observation of any difficulties. None of this information advanced matters appreciably and I therefore sought further information both from BNFL and UKAEA. As a result Mr B F Warner of BNFL reviewed the available evidence and reached the conclusions that:—
  a. It is probable that zircaloy fuel may be stored for up to 20 years, and remain suitable for handling and reprocessing.
  b. It would be imprudent to store substantial quantities of stainless steel clad fuel in ponds for more than a decade.
  c. Further evidence was required before present plans for early reprocessing could prudently be modified.

8.15 In addition, BNFL accelerated a research programme upon which they had already embarked and UKAEA undertook certain further researches. The results of the accelerated and further work were given in evidence by Dr R H Flowers, the head of the Chemical Technology Division at the UKAEA Harwell laboratory on the 86th day of the Inquiry, and I should like to express my thanks for the work done. The conclusions reached were necessarily based on a limited programme but so far as they went they confirmed Mr Warner's conclusions set out above.

8.16 Tests on fuel pins from the Windscale prototype AGR had revealed corrosion processes capable of penetrating the cladding after 3–4 years in the pond. The processes were of two types. In fuel with a low (400–550°C) in-reactor temperature range it consisted mainly in corrosion of sensitised grain boundaries although there was also evidence of more general attack. In fuel with a high (600–750°C) in-reactor temperature range there was little evidence of grain boundary attack. In this temperature range there was however a marked loss of wall thickness, leading to some penetration. Windscale AGR fuel cladding is 250 μm thick whereas commercial AGR fuel cladding is 375 μm thick. Dr Flowers did not feel able to make any confident prediction of satisfactory pond life of such fuel but considered it would be wise to assume that some local penetrations would occur after five years unless the water treatment was different from that presently employed. This period might be extended but he was unable to say for how long. Zircaloy clad fuel with pond life up to nine years was also tested. No evidence of corrosive attack had been found up to the time when Dr Flowers gave evidence. In the case of such fuel Dr Flowers felt unable to predict maximum storage life but considered that it might be more than 20 years.

8.17 At first sight it may appear surprising that little or no research has been done into the question of the satisfactory pond storage life of either form of spent fuel or into the possibility of prolonged storage by other means, but this is in reality not so. Until recently it was the common assumption that reprocessing would take place after a relatively short period of pond storage and attention had been directed to ensuring that the fuel pins would stand up to reactor service rather than to their ability to withstand pond life.

8.18 The general situation is well stated in two published documents put in evidence by FOE which I quote:
  1. 'In the uranium fuel cycle initially conceived for the light water reactors (LWR), spent fuel was to be discharged from the reactors and then allowed to cool (at the reactor sites) for five months. This cooling period would allow short-lived radioative isotopes to decay and thereby reduce the heat generated. After the cooling period, the fuel was to be shipped to a chemical reprocessing plant, which would recover uranium and plutonium. Accordingly, existing storage facilities at LWR sites were designed for only short-term storage of spent fuel. More recently, however, spent fuel has been recognised as a possible waste form suitable for interim storage and even ultimate disposal'. (FOE 68)
  2. 'Reprocessing has for so long been a part of plans to complete the nuclear fuel cycle that little attention has been paid to ways in which a nuclear power system might operate without it'. (FOE 66)

8.19 On the evidence before me it is clear that, if there is going to be a delay in the commencement of reprocessing of AGR fuel, an urgent research programme is necessary to determine whether the cladding can be so designed, or pond storage methods so adjusted and improved, as to make increased pond storage life prudent. That it is not prudent with existing design methods I have no doubt. Research and development will also be necessary into storage for periods of 50 years or more for, if reprocessing is delayed now, there can be no certainty that it will not be delayed again. Canada for example is considering a variety of options for storage up to 75 years in retrievable storage so that the 'fuel could be recovered for reuse whenever the economics become favourable' (FOE 67). Canada, however, is, like the US, favourably placed for uranium supplies and does not use stainless steel cladding. In a letter to FOE dated the 2 August 1977 (UKAEA 3) Mr W Morgan of Atomic Energy of Canada Ltd wrote (stressing that it was a personal opinion) 'I am convinced that re-cycle of fissile material is essential. Thus reprocessing is essential and the sooner we (the world) get on with it, the better. At the moment a special situation applies in Canada. We have large per capita uranium reserves and the extent of as yet undiscovered resources is also thought to be appreciable. Thus we are under no immediate pressure to re-cycle fissile material. So interim fuel storage looks attractive and this accounts for our work on the subject'.

8.20 If it were now decided to reprocess, but to delay the start for 10 years, it might be that the research would be of a limited nature and that additional life in ponds could be rendered satisfactory comparatively easily. Since, however, no such decision is advocated by anyone and is in any event not practicable, it appears to me that, if the decision is 'not now', BNFL must urgently develop and instal long-term storage facilities to which fuel could be transferred after an initial period in ponds. That they could do so is not in issue. BNFL readily accepted that they could but the development would take many years. It is by reason of the different direction of effort that delay would involve, that they contend that a firm decision must be reached at this stage. Whether BNFL should be committed to the research, development and construction of long-term storage facilities necessarily involves some assessment of the merits and probabilities of the alternative ultimate disposal routes. To this question I now turn.

8.21 Suggestions made at the Inquiry included the following (with some variations)
1. Indefinite storage with the fuel being moved at prolonged intervals to newly constructed storage facilities.
2. Disposal of the spent fuel as such after an extended period of storage.
3. Disposal of resulting waste after a form of reprocessing involving separation of uranium but not plutonium.
4. Disposal of resulting waste after separation of plutonium and uranium as presently contemplated.

8.22 Alternative 1 above appears to me to have little merit. It was not seriously pressed by anyone and I reject it. The main merits of alternative 2 are the non-separation of plutonium and the avoidance of the discharges of radioactivity involved in reprocessing. Its disadvantages lie chiefly in the wastage of the energy locked up in the plutonium and uranium and the fact that all the plutonium, with its very long radioactive half-life, could escape from the disposal site back to the environment. Alternative 3 avoids the wastage of uranium but would probably involve discharges in reprocessing and still commits all the plutonium to potential return to man. Alternative 4 avoids these disadvantages but involves discharges to the environment and raises the problem of disposing of the plutonium.

8.23 It was common ground that, by reason of the very long periods during which the contents of the spent fuel would remain radioactive, planning for ultimate disposal must proceed on the basis that containers would or might be breached, that water would or might get to the spent fuel whether it was in wholly unreprocessed form or not and that, if water did get to it, leaching might occur which could lead to its return to man. The risk of this happening was considered by some to be such that nuclear power should be abandoned altogether and at once. That is another question. For present purposes it does not arise. The questions of prime importance are (i) whether plutonium should be separated; (ii) whether the highly active waste can be put into a form suitable for disposal, which, it was common ground, requires solidification, and (iii) whether when solidified it can be safely disposed of.

8.24 The merit of reprocessing in the present context is that it extracts almost all the long lived plutonium from the waste and thereby leaves much less to return to the environment from the disposal site. That this merit is considerable is easily demonstrated. If spent fuel with a content of 100 tonnes of plutonium 239 is committed to final disposal, it will be 244,000 years before the plutonium 239 has, by decay, reduced in quantity to 0.1 tonne. If, however, that same quantity of spent fuel were reprocessed there would be only 0.1 tonnes of plutonium 239 remaining in the waste to be sent to the final disposal site in the first place. In the period of 244,000 years that amount would by decay have reduced to about 100 grammes. Our responsibilities to future generations therefore appear to demand reprocessing unless the plutonium extracted (or an equivalent quantity) must itself be sent to final disposal at some later date.

8.25 In the early stages of the Inquiry it was assumed that all separated plutonium (or an equivalent quantity) would ultimately need to be disposed of. This, however, is not the case. Plutonium can be used in mixed oxide fuel for thermal reactors and if it it so used the total plutonium inventory can be thereby reduced.

8.26 For the purposes of illustration I assume (1) that a number of AGRs using uranium fuel are operated at 70 per cent load factor for a period sufficiently long for the plutonium content in the spent fuel arisings to total 12 tonnes and (2) that stocks of magnox plutonium amounted to 55 tonnes at the commencement of the period. The total plutonium inventory at the end of the period would then be 67 tonnes, of which 55 tonnes would be separated and 12 tonnes unseparated. If, however, the same AGRs operating at the same load factor and producing the same amount of electricity were fuelled with mixed oxide fuel it is estimated that about 46 tonnes of plutonium stocks could be fed to the reactors of which only about 32 tonnes would remain in the spent fuel. In this case, the total inventory at the end of the period would be 41 tonnes, of which 9 tonnes would be separated plutonium and 32 unseparated. The result of reprocessing and use in the AGRs would thus have resulted in there being about 26 tonnes less plutonium to be disposed of than if no reprocessing had taken place. The above figures are estimated to be the result of running reactors with a production totalling 10 GW(E) per annum for about $5\frac{1}{2}$ years. It will thus be seen that if reprocessing takes place there can be, on a single re-cycle, a much reduced total inventory. The price of the reduction is of course the discharge of radioactivity involved in the reprocessing. This I examine later. What it is necessary to note here is that there will be substantially more plutonium against which to protect future generations by not reprocessing than by reprocessing.

8.27 There is also to be considered the comparative ability to resist leaching by water of the spent fuel and the types of glass under consideration for the vitrification disposal route if water reaches them. The evidence indicated that the leach rates per unit of surface area of fuel pellets and glass were much the same, but the surface area of the fuel pellets would be many times that of the glass required to contain the highly active waste resulting from reprocessing. Moreover there is at least a possibility that the UO_2 content of the fuel pellets would oxidise to U_3O_8. This would involve swelling and fragmentation of the pellets with the resultant exposure of even greater surface areas to leaching action.

8.28 Such leaching action in the case of plutonium could be particularly serious for not only might the plutonium get back to man as such but it might accumulate, for example on a clay deposit, reach a critical mass and then release highly active fission products.

8.29 For the objectors, the three questions mentioned at the end of paragraph 8.23 above were principally dealt with by Professor Tolstoy, scientist, university professor and writer giving evidence on behalf of WA. The evidence of BNFL's witnesses was supported by that of Dr Stanley Hay Bowie, FRS, assistant director of the Institute of Geological Sciences from 1968 until his retirement during the course of the Inquiry, who was called on behalf of Ridgeway Consultants.

8.30 Professor Tolstoy stressed that neither the HARVEST process for vitrification nor any other process had yet been finally proved. This is correct. On the other hand BNFL's witnesses were confident that the feasibility of vitrification had been established. What remained was to optimise and demonstrate it on a large scale. I accept the evidence that a vitrification process will be successfully established. The stage of development has, I consider, been reached, when success can be confidently predicted. Indeed success must, either by the HARVEST process or some other process, be achieved. Much highly active waste has already been accumulated and there is no option but to continue to reprocess magnox fuel. By the end of the useful lives of the magnox reactors there will be accumulated large additional quantities of highly active waste. All of this can only be rendered fit for disposal by some form of solidification process.*

8.31 On the question of disposal of the solidified waste Professor Tolstoy drew attention to a large number of points which showed that a final solution to the problems of disposal has not yet been found. This I accept. It is however equally true that no solution has been found to the problems of disposing of spent fuel if it is not reprocessed and many of the points raised by Professor Tolstoy applied equally and indeed with more force to the disposal of spent fuel. At the conclusion of his evidence he very fairly accepted that it was desirable to have as little plutonium in the waste as possible, and that as between glass blocks and fuel pellets the fuel pellets would have much greater surface area exposed to leaching. He also relied on an article in the April 1977 issue of 'Metals and Materials' (WA 139) reporting a critical assessment by Dr G Wranglen, Professor of Corrosion Technology at Sweden's Royal College of Technology, Stockholm, of a Swedish Government Report. In that article there appears the following:
> 'It has recently been proposed that spent radioactive fuel, rather than high level waste, should be disposed of without reprocessing by burial in the earth. This is quite inadmissible, states Dr Wranglen.'

Professor Tolstoy's final remark on the subject was:
> 'May I say that I certainly am not proposing, nor are we proposing anywhere, that the spent fuel be disposed of without any form of treatment.'

8.32 The final effect of Professor Tolstoy's evidence was to confirm that, as between disposal of spent fuel and solidified highly active waste the latter was the preferred method since it would involve the disposal of much less plutonium and be less vulnerable to leaching. In this he was at one with BNFL's witnesses and Dr Bowie. His real

---
*The possibility of separating and subsequently incinerating in reactors the actinides contained in the highly active waste, a matter which was considered briefly in the Sixth Report (paras 384–387) was not canvassed at the Inquiry.

objection was not to reprocessing in THORP as such, but to the great increase in waste – be it in the form of unreprocessed spent fuel or in the form of highly active waste resulting from reprocessing – which he considered would follow after THORP if THORP were built. He was thereby taking the view adopted by the Royal Commission that there should be no commitment to a large programme of nuclear fission power until waste disposal problems had been solved. THORP would not in my view involve such a commitment for reasons which I have already expressed. I add only that, quite apart from those reasons, the Government's express assurance during the Inquiry that there would be a further Inquiry before any decision to build CFR 1 was made, confirms that the Government do not regard THORP as involving any such commitment.

8.33 A question repeatedly posed at the Inquiry was 'Have we the right to expose future generations' to one risk or another. In this connection the appropriate question might be, 'Have we the right to expose future generations to the possible escape of much more plutonium than is necessary and in a form which is more vulnerable to leaching than it need be?' I cannot of course answer that question but I can say that, given the option to commit to the 'nuclear dustbin' considerably less plutonium, in a form less vulnerable to escape than would be the case if there was no reprocessing, I would need some compelling reason to make me feel it 'right' to opt for the greater quantity in the more vulnerable form.

8.34 The next aspect of need is that of energy conservation and resource independence. It is to be observed that neither the USA nor Canada are presently troubled by the latter problem in respect of uranium supplies. Both have uranium supplies sufficient not only to supply their own needs for many years but also to allow them to use the threat of withholding supplies as an instrument of policy, the effectiveness of which is beyond doubt. In his final speech Counsel for FOE accepted that such a threat could be used not only to force another country not to start reprocessing but also to enforce the acceptance of a policy which that other country considered morally abhorrent. It appears to me that if we are going to depend, for a substantial part of electricity supplies, upon nuclear power, it is in the public interest that we should, unless the price of so doing is too great, minimise reliance on imported fuel.

8.35 If oxide fuel reprocessing is undertaken the energy which can be produced from a given original quantity of uranium will be increased. The amount of the increase will vary according to the type of reactor in which recovered uranium and plutonium are used. If used in AGRs, recovered uranium alone would produce about 15 per cent more energy. Recovered uranium and plutonium together would raise the increase to 30 to 40 per cent. In FBRs the increase would be very great indeed. It could reach as much as 60 fold. To dispose forever of spent fuel before we are sure that the energy locked up in it will not be required to meet our own needs or, by releasing supplies of uranium, the needs of others, would therefore be an act of folly. This does not however require a start on THORP now. Such a start would only be needed from an energy viewpoint if in 10 years time there might be a need for plutonium derived from oxide fuel reprocessing. I have already concluded that there would not. Magnox plutonium will suffice well beyond that time.

8.36 The resource independence aspect is however different. If THORP is begun without delay it can be adding to indigenous supplies by 1990 or thereabouts and could produce by 2000 another 17 tonnes of plutonium together with approximately 4,200 tonnes of slightly enriched uranium*. The question which therefore arises is this. Would it be prudent for this country to have such additional supplies available then? This must depend to a great extent upon the question of energy needs and the possible means of meeting such needs.

8.37 Forecasts of energy demands which were advanced covered a very wide range as did predictions as to how such demands could or should be met. They were made against a background of decreasing forecasts as to needs, increasing attention being paid to conservation measures and possible alternative sources of energy, the relief given by the availability for a time of oil and gas from the North Sea, the very large reserves of coal which exist in this country and the existence of world supplies of uranium which would probably be sufficient to fuel the world's thermal reactors well past the turn of the century. Nevertheless it is clear that such evidence fell far short of what I would require were it for me to make a definitive forecast. I have not regarded it as any part of my task to attempt to do so. It would serve no useful purpose, for, no matter how much evidence had been tendered, any forecast which I might make would be as uncertain as any other forecast. As was pointed out in the course of this Inquiry the only certain thing we know about the future is that we do not know it. Any planning for the future must therefore be continually revised to take account of events which occur. North Sea operations may suffer mishaps which will result in supplies running out far sooner than presently expected, or further exploitable reserves may be found which will extend the time for which they may be available. The development of one or more alternative energy sources may turn out to be more or less promising than presently appears. Some may fail altogether. Others may be capable of producing more energy than presently predicted. The public may respond well or poorly to conservation measures. An alternative energy source, whilst proving technically feasible, may have great environmental drawbacks. And so on. The list is endless. With so many uncertainties the only prudent course is to adopt a strategy which will give the greatest assurance that, no matter how the variables change, the energy needed to support an acceptable society can be provided.

8.38 The foregoing may all appear too self evident to need saying. I say it because some parties and witnesses appeared completely to overlook the fact that there is a

---

*i.e. from reactors presently existing or under construction.

great difference between, on the one hand, making a confident forecast, without having either the power to act upon it or any responsibility for the consequences if, when someone else has acted upon it, it proves to be wrong, and, on the other hand, taking and acting upon a decision the consequences of which will affect the lives of millions. It is the Government which has the power and the duty to make such decisions.

8.39 The latest (October 1977) set of forecasts put in evidence was that contained in Energy Commission Paper No 1 (G 70) prepared by the Department of Energy for the recently created Energy Commission. The substance of this had, however, previously been given in evidence by Department of Energy witnesses. Expressed in millions of tons of coal equivalent (mtce), the forecasts for total primary fuel demand for the year 2000, for alternative assumptions as to growth of the economy, were 560 for the high growth assumption and 450 for the lower assumption. These figures may be compared with figures of 650 for the top and 500 for the bottom of the range contained in the Energy Policy Review (G 3) prepared by the same Department in January 1977. The contribution of nuclear power to those needs was, in the earlier document, envisaged as being possibly as high as 100 mtce and in the latter 95 mtce. When considering the figures it should be noted that reactors presently in existence or under construction together with those currently planned would account for only about 30 mtce.

8.40 The latest forecast of total energy demand may well have to be reduced again and, whatever the forecast of total demand may be, the expected contributions from various sources may change. But none of the evidence given led me to believe that it would be otherwise than imprudent not to continue to develop nuclear technology and keep the nuclear industry in a condition to meet a sudden expansion in nuclear power should it be required, be that expansion in thermal reactors or in FBRs. As is well recognised by the Government, efforts to effect energy savings by conservation methods, such as insulation or combined heat and power schemes and to develop alternative sources of energy such as solar, tidal, wind, wave, biomass or geothermal should certainly be pursued; but to divert available resources to such efforts to an extent which would prejudice a large scale reliance on nuclear power should it be needed would, it seems to me, be an act of bad management for which this and future generations might justly blame the Government should such reliance prove to be required. Much was made of the ability of coal reserves to deal with any shortfall from other sources, in particular by Mr Arthur Scargill the President of the National Union of Mineworkers (Yorkshire Area) who declared himself a passionate opponent of nuclear power. I can only describe his forecasts of what could be achieved as fanciful. Moreover, even if he were right in his forecasts of what could be done, it by no means follows that increasing the capacity of coal fired stations instead of nuclear stations would be of benefit to this or future generations. Four quotations from 'Nuclear Power Issues and Choices', the report of a Nuclear Energy Policy Study Group sponsored by the Ford Foundation (The Ford Foundation Report – BNFL 39) are of interest in this connection.

a. 'In the case of coal, several hundred years of experience have not produced quantitative understanding of the health consequences and even less understanding of the possible effects on the world's climate of the carbon dioxide and particulates released during coal combustion'. (P 16.)
b. 'Despite these large uncertainties, the general conclusion is that on the average new coal-fueled power plants meeting new source standards will probably exact a considerably higher cost in life and health than new nuclear plants. . . . The most pressing demand, however, would appear to lie in upgrading the research and development directed at the reduction of the adverse health effects associated with coal-fueled power plants'. (P 196.)
c. 'The possible impact on global climate appears to be the most serious environmental consequence of greatly increased electric power generation. The thermal output of both coal and nuclear power contribute directly to the long-term heating of the atmosphere. However, a much more serious threat appears to be posed by the carbon dioxide (CO_2) produced in fossil fuel combustion'. (P 210.)
d. 'On balance, the local environmental consequences of the nuclear power cycle in normal operation are not as serious as those from fossil fuel power generation. These local effects, however, are less critical in an overall evaluation of potential environmental impacts than the effects of increased greenhouse heating on global climate. While it is not possible at this time to judge the nature of the potential impact on civilization, it may develop on the basis of greater knowledge that this global climatic effect could be overriding in a comparison of coal and nuclear power. This argues against putting complete reliance on coal power at this time'. (P 211.)

8.41 The emissions from coal burning stations which were considered when making the statements quoted included sulphur dioxide, carbon dioxide, nitrogen oxides, carcinogenic hydrocarbons, radioactive and other particulates and heavy metals. It is not specifically so stated but the heavy metals would include lead and mercury, both of which are highly toxic and both of which are stable and thus persist for very long periods in the environment. Mutagens were not mentioned but I had independent evidence that coal burning results in releases of mutagens.

8.42 The above matters are also relevant to risks but I have felt it desirable to include them under the heading of need. Much stress was laid on the large coal reserves available in this country and at first sight the argument that there is no need to expose ourselves to the risks of nuclear power when there is 'all that coal' waiting to be won is attractive. 'That coal', however, carries its own risks which, quite apart from the risks from mining, which might be described as voluntarily accepted by miners, are every bit as much imposed as are the risks from nuclear plant. Indeed such risks also include risks from radioactive emissions, a fact which is not widely known.

8.43 Reverting to the matter of forecasts, Dr Peter Chapman, Director of the Energy Research Group of the Open University, who was one of the principal witnesses for FOE, envisaged for the year 2025 a nuclear contribution to energy needs at that time of some 25 GW(E) of installed capacity. This forecast is below the Department of Energy forecast but it nevertheless involves the creation by 2025 of some 16 GW of extra nuclear capacity. Dr Chapman envisaged a steady increase from about 1985 onwards. His evidence was impressive for, whilst critical of other forecasts which had been made, it was characterised by a moderation and rational argument notably lacking in some of the evidence tendered. I quote a passage which appears to me to be of considerable importance.

> '... the rate of growth in the nuclear component is seen as steady, and I would regard that as essential and important. Part of what I have been saying is that I want to maintain the capacity to increase the supply capacity of each of these fuels, that is, nuclear, wave, solar and so on. So I want to maintain a nuclear industry, an industry capable of building nuclear power stations. So I would want to have a steady ordering programme throughout that period to ensure that they had the capacity to build nuclear power stations.
> Q. Yes. So that assuming you have got another 16 gigawatts to fit in and assuming that your stations are, let us say, 1.2, broadly speaking, you would divide the period by the number of extra stations and assume that they come on line regularly during the period?—A. Something along those lines sir, and there would also of course be retirements to be taken into account. Certainly the magnox stations are expected to be coming to the end of their useful life-time about the turn of the century and the AGRs presumably would need to be replaced before 2025 as well.
> Q. So that you would in fact be building to get up to your 25 gigawatts, you would be building 16 gigawatts of additional new capacity and towards the end of that period you would also have had to start replacing magnox and early AGRs possibly?—A. Yes.
> Q. Is that right?—A. Yes, sir.'

8.44 This passage recognises the fact that, if we are to preserve the capability to expand nuclear power, the industry has got to be preserved. Assuming an additional 16 GW(E) extra over the 40 year period from 1985 to 2025 this would, on a regular basis mean no more than one 1.25 GW(E) station every 3 years but, as Dr Chapman pointed out, the magnox stations and the earlier AGRs would need replacing in the period. If one assumes replacement of present capacity within the period i.e. the building between 1985 and 2025 of 25 GW(E) capacity, this would give an average rate over the period of one station coming on line every two years. Assuming such a programme there would be an additional 8 stations of 1.25 GW(E) by 2000, all of which would be producing spent oxide fuel.

8.45 Dr E P Radford, Professor of Environmental Epidemiology at the University of Pittsburgh, a witness for the Network for Nuclear Concern (NNC) who had long been and still remained, a supporter of nuclear energy development stated his reasons for such support:

> 'I believe that we are going to need every resource available to bring to bear on the energy crisis that has been really quite obvious now for the last 15 years. In other words the dwindling supplies of fossil fuels has signalled the fact that we will have to substitute other sources of energy or else convert to a relatively low energy lifestyle which I do not accept and I do not believe the vast majority of people accept as being an adequate basis for life, especially with the world population being what it is.'

8.46 Mr Peter Taylor of the Oxford Political Ecology Research Group (PERG) considered that an expansion of nuclear power was the only way of maintaining a growth economy.

8.47 The TUC submitted written evidence in which it expressed a similar view. It envisaged, as a result of a number of reviews, that a major contribution from nuclear energy would be needed by the year 2000 and at its Blackpool conference in September 1977 passed a resolution which included the following:

> 'Congress instructs the General Council to press the Government to formulate a plan for energy.... This should include the following objectives....
> (ii) To maximise the contribution of an expanded and socially acceptable nuclear programme which is consistent with the maintenance of a safe environment in terms of solving problems of health and security which may arise.'

8.48 It appears to me that, if the nuclear industry is to be kept in being, the sort of programme envisaged by Dr Chapman would be about the minimum necessary to achieve this; indeed it is significant that his programme was envisaged, notwithstanding that he regarded it as feasible that the capacity alloted to nuclear power could be supplied by waves, wind, hydro and tidal power and that he assumed, over and above this, an annual contribution from solar energy of 38 mtce.

8.49 If a programme of such a size is necessary for such a purpose then, unless the option to expand nuclear power even further is to be jettisoned, and I had no evidence which convinced me that from an energy outlook this was desirable or reasonable, planning must be on the basis that oxide spent fuel arisings from thermal reactors might be as much as 600 tonnes per year by

2000. If they were and, if we are to reprocess, we should then need a plant of sufficient capacity not only to process the accumulated backlog which would exceed 6000 tonnes but also to deal with such annual arisings and increasing annual arisings thereafter. And we should need this whether or not the nuclear contribution could then be met from alternative sources.

8.50 In the light of the foregoing I propose to deal with conservation measures and alternative energy sources very briefly. Their importance, and the devotion of increased effort to them, is, as I have said, well recognised by the Government. It is also recognised generally. I refer to Energy Commission Paper No 1 (G 70), the Third Report from the House of Commons Committee on Science and Technology (BNFL 275), the TUC submission and resolution already referred to (TUC 2 & 3), the Sixth Report (BNFL 9), the White paper (BNFL 170), and the Watt Committee Working Party's first report (TRIB 12). The evidence of objectors and applicants alike was to the same effect. The area of dispute was confined to the contribution to energy needs which might be expected from such measures and alternatives.

8.51 As to this, I repeat that none of the suggested measures or sources appeared to me sufficiently likely, either alone or in combination, to produce such savings in energy or contributions to energy supply as would justify or render reasonable any course other than a steady development of nuclear power. Moreover it is necessary to remember that technical capability is not the only thing which is of significance. For example, the installation and maintenance of strings of oscillating generators $1\frac{1}{2}$ miles off-shore to the north of Scotland or the erection of 4,800 windmills in the North Sea would create dangers of their own not only to those who might build and maintain them but also to shipping. Another possibility, the planting of the Lake District with thousands of acres of sugar beet or eucalyptus trees, which was discussed at the Inquiry, would be unlikely to receive much of a welcome from environmentalists.

8.52 The fact that I do not rehearse in detail the various possibilities spoken to in evidence must not however be taken in any way to mean that I regard the evidence given as unimportant or the suggestions as being unreal. The witnesses who spoke of them were, on the whole, moderate, but because the contributions of which they spoke are of necessity uncertain, it does not seem to me that any detailed consideration would be of assistance for present purposes.

8.53 It was urged in argument that, if more funds were devoted to research into and development of conservation and alternative sources of energy, then their contributions would be greater; and that the devotion of funds to nuclear development is wholly disproportionate and prevents such other contributions being increased. In this connection it must be pointed out, although it may be thought somewhat elementary, that the funds which are needed for, and capable of absorption in, a research and development programme for a particular project, increase as the programme proceeds and that, if a particular project has proceeded to a stage of near success, (i) the final stages will probably require very large funding and (ii) to direct funds, when success is near, to another project which is a long way back on the development road is hardly sensible. This inevitably means that an ambitious project achieves a momentum of its own. A stage will or may be reached when so much has been spent on it, and so many jobs are dependent on its continuation, that it is difficult to stop. This is however not a sound argument for stopping, or for failing to embark on, ambitious projects. It is a sound argument that the fact of momentum must be recognised and care taken to see that the momentum is not allowed to override other considerations. Thus, if the risks inherent in an increase in nuclear power went beyond a certain point, the momentum should not be allowed to prevail. If on the other hand the momentum, if allowed to prevail, would mean no more than some delay in the progress of some other project this would not alone be enough. Save where resources are limitless the final development of a large project will always lead to delays in others and, if this were not acceptable, no large project would ever reach fruition.

8.54 Assuming that a reprocessing plant is needed, the question which next arises is 'How large should it be?' with which is bound up the matter of foreign fuel reprocessing. For a time there can be no doubt that, purely to meet our own reprocessing needs, a 600 tonne plant would be sufficient. BNFL do not suggest otherwise. But that there should be capacity to reprocess fuel for other countries, is, as I have already said, desirable. It will become more so as time passes, and if, as I think, the taking of foreign business will relieve the pressure on other countries to develop reprocessing capabilities for themselves, to have a plant large enough to do so and to use it for such business is desirable on that ground, unless considerations of risk lead to a contrary view. It is even more desirable if, by taking the foreign business, the ultimate cost to the consumer can be reduced and foreign currency earned. These are matters which I consider in the next section.

8.55 On the question of need it is convenient at this stage to summarise my conclusions. They are as follows:
  a. Oxide fuel reprocessing is not necessary for the purposes of preserving the option to build CFR 1 or for the purposes of enabling an FBR programme to be launched. Plutonium from magnox reprocessing can provide sufficient for both purposes.
  b. To develop an FBR programme beyond the 8th FBR would require oxide fuel reprocessing and would also probably require additional thermal reactors or a much slower introduction of FBRs.
  c. Additional plutonium from oxide fuel reprocessing will not be required until after 1987. Purely from the point of view of plutonium production a start on oxide fuel reprocessing could therefore be delayed. This delay could be for at least five years

and possibly for 10 or even more years.
d. If there were delay, but reprocessing were ultimately required, it would be necessary to reprocess at a greater rate and in a plant of larger capacity than is presently required. In the intervening period experience in reprocessing and the results of the discharges involved would not have been gained, save in so far as operation of B204/205 provided such experience. This would be of lesser value than experience on a new plant working on a regular production basis. As a consequence risks would be greater when a start was ultimately made.
e. Delay would also involve storage of spent fuel for longer periods. It would be imprudent to store the present type of AGR spent fuel for any substantial additional period. Delay would therefore require an urgent programme of research and development into methods of storage for spent fuel for longer periods. Such periods would have to be as long as 50 years or more since, if there is a delay now, there can be no guarantee that there will not be further delay.
f. It is undesirable to dispose of spent fuel without reprocessing and extraction of the plutonium because to do so (i) involves committing future generations to the risk of the escape of more plutonium, for longer periods and in a form more vulnerable to leaching (and thus escape) than would be the case if reprocessing were conducted; (ii) involves throwing away resources which could provide energy over long periods.
g. There is little to be gained from research and development into long term storage because it is undesirable to dispose of spent fuel without reprocessing.
h. It is prudent to keep the nuclear industry in being and able to build further reactors in case of need. This involves at the least a regular building programme of thermal reactors and thus increasing quantities of spent fuel arisings.
i. The possible need for a considerable nuclear programme may include the need to reduce reliance on coal since coal fired stations are probably at present a greater source of harm than nuclear powered stations.
j. There is a world need now for reliable reprocessing capacity.

8.56 In the light of the foregoing I conclude that reprocessing involving extraction of plutonium is desirable and will be required at some time. I further conclude that if it is to be required at some time there should be no delay in building the plant.

9. Financial aspects

9.1 Although BNFL made alleged financial advantages part of their case, no detailed financial analysis was produced by them and for this lack they were, in my view rightly, criticised by more than one objector. It was indeed submitted that unless and until the alleged financial advantages had been established by such a detailed analysis no consent should be given. Whilst I accept the criticism I do not accept the submission. I reject it for two reasons:
  i. because it would be a serious defect in the procedure if a party, who had made his case out on the evidence, were to have his application disallowed because that evidence had not taken a particular form, adherence to which was not required by law;
  ii. because even if, in the absence of proof as to financial advantage, financial disadvantage must be assumed – which I do not accept – it might nevertheless be right to grant permission. The financial disadvantage might be an acceptable price for some other advantage, for example resource independence, reduction of plutonium stocks, or anti-proliferation effect.

9.2 BNFL's case as initially presented was very brief. Mr Conningsby Allday, their Managing Director, pointed to the resource savings which would flow from reprocessing, quoting by way of example the number of fast breeder reactors which could be fuelled for their life times from the plutonium and uranium recovered from reprocessing without there being any need for further imports of uranium. He then went on:—
> 'The economic benefits of reprocessing derive directly from the resource saving arguments set out above. Meaningful comparison of costs for comparable reactor programmes, with and without reprocessing and subsequent plutonium and uranium recycle, is however made difficult by the fact that no waste management route other than reprocessing has been developed to the point where cost estimates can be made on the same basis of experience as for the reprocessing route'.

9.3 The fact that there would be resource savings does not, however, necessarily lead to economic benefit. If, for example, the costs of separating uranium exceed the costs of obtaining, via imports, a like quantity of oxide, enriched to the same extent as the recovered uranium, it will be cheaper, if one considers only the question of uranium supplies, to adopt the import route (I leave plutonium out of account for the present, purely for the sake of simplicity). To make such a comparison is however as defective in providing useful information as it is to look at resource savings only for, if spent fuel arises and is not reprocessed, something else must be done with it.

9.4 The matter having been opened up in the early stages of the Inquiry, discussions took place between representatives of BNFL, the CEGB and objectors with a view to providing me with information which would enable some useful assessment of the financial implications to be made. These discussions resulted in the production of two documents dated respectively the 19 August and 6 September 1977 (BNFL 232 and 265) being respectively the 48th and the 59th days of the hearing. These documents, together with certain other documents and the oral evidence given on the financial side, principally by Dr Chapman for FOE and Dr Sweet for WA, afford, in my view, ample material for the type of assessment which can reasonably be made at the present stage.

9.5 From the evidence produced it is clear that, viewed simply as a plant for the production of uranium and plutonium, the plant would not break even unless and until the price of uranium increased very markedly against reprocessing costs. At present reprocessing costs, the break even point would be when uranium prices advanced to about \$80 per lb, or more than twice the present level. A doubling in price by the year 2000 can well be contemplated but, by then, reprocessing costs too would have risen. Viewed purely as a production plant it appears to me that THORP would be economically disadvantageous in absolute terms. Whether the disadvantage in such terms would or would not be outweighed by other considerations would then fall to be considered although that would not, as I see it, be a planning matter.

9.6 Since, however, the spent fuel, if not reprocessed, has to be dealt with in some way, the costs of so dealing with it cannot legitimately be ignored. To take them into account is necessarily speculative, for no real work has been done on any waste disposal route other than reprocessing. Certainly there would have to be further storage facilities, which might be either 'wet', i.e. cooling ponds, or 'dry', that is to say storage in an inert gas environment. At some stage the fuel would have to be prepared for ultimate disposal.

Such preparation might involve no more than handling facilities and the encasement of the fuel in suitable containers, or it might involve some form of reprocessing followed by vitrification, or other means of converting the highly active waste into a suitable form for ultimate disposal. Such reprocessing might be as presently contemplated, for in twenty or thirty years time it might be found necessary or desirable to recover both the uranium and the plutonium, even if it is not necessary or desirable now. It might be that it was in a more limited form, involving recovery of the uranium only, or it might be that it consisted merely in the dissolving of the fuel. With such possibilities, none of which have been researched, any estimate of the costs which will be incurred if there is not to be reprocessing can only be very rough. Such costs must clearly include the necessary research and development into methods of storing spent fuel satisfactorily for long periods, the provision of such facilities and the costs of running and maintaining them. BNFL have estimated that, assuming storage to the year 2036/37 the price to the home boards would be some £225,000 per tonne uranium for the 'dry' storage alternative and some £150,000 per tonne uranium for the 'wet' storage alternative, as against an estimated price for reprocessing and vitification of £260,000 per tonne for reprocessing in THORP or £315,000 per tonne if a smaller plant capable of dealing only with home arisings and existing foreign commitments were built.

9.7 If the above figures are right, reprocesssing has an immediate advantage compared with the dry storage.

The recovered uranium will be worth, assuming a price of $30 per lb, some £60,000 to the customer. The net cost to him, of reprocessing through THORP will thus be some £200,000 to compare with storage of £225,000. It would be immediately disadvantageous to the extent of some £50,000 per tonne if the wet storage option proved feasible.

9.8 In the light of the evidence on corrosion I do not consider the wet storage option would be likely in the case of AGR fuel. Hence the plant would, on BNFL figures, be advantageous even if no account is taken of costs from storage to preparation for ultimate disposal. On the other hand the smaller plant mentioned above would be presently disadvantageous, for after allowing for recovered uranium the net cost would be £255,000 per tonne against £225,000 per tonne for dry storage.

9.9 If, however, reprocessing of some sort, followed by vitrification, were added there would be a clear advantage to reprocessing. BNFL estimate the cost on such basis, taking credit for recovered uranium, would be from £615,000 per tonne (at a uranium price of $30 per lb) to £475,000 (at a uranium price of $100 per lb).

9.10 I accept BNFL's evidence that some form of reprocessing followed by vitirfication is the most likely route to ultimate disposal and I conclude therefore that the plant is likely to be advantageous financially, whether it is of the capacity presently proposed or of a reduced capacity, unless BNFL's estimates are very seriously at fault.

Table Comparative Costs to Home Boards of Reprocessing at Uranium Price $30 per lb (Sources BNFL 232 and 265)

	Cost of reprocessing including vitrification of waste		Cost of long-term storage only		Long-term storage followed by reprocessing to recover uranium and vitrification of waste (after credit for recovered uranium)
	Using THORP as proposed	Using smaller plant dealing with home arisings and existing foreign commitments only	Wet storage	Dry storage	
	£ per tonne	£ per tonne	£ per tonne	£ per tonne	£ per tonne
Initial cost	260,000	315,000			
Credit for recovered uranium	60,000	60,000			
Net	200,000	255,000	150,000	225,000	615,000

Notes:
1. Net cost of reprocessing in THORP followed by virtification is higher than cost of 'wet' storage but lower than cost of 'dry' storage (para 9.7).
2. Net cost of reprocessing in smaller plant followed by virtification is higher than cost of either 'wet' or 'dry' storage (para 9.8).
3. Net cost of reprocessing in either THORP or smaller plant is much less than cost of storage followed by limited reprocessing and vitrification (para 9.9).

9.11 The main attack on BNFL's figures is to be found in a memorandum submitted by Dr Chapman in October 1977 (FOE 130). This was submitted long after Dr Chapman had given evidence and he was not cross-examined upon it. This was no fault of Dr Chapman's or of FOE who called him, for BNFL's two documents to which I have referred were submitted too late for them to be considered and dealt with any earlier. In answer to Dr Chapman's memorandum BNFL submitted written comments (BNFL 327) on the 98th day of the Inquiry after all objectors had made their closing submissions. There was no cross-examination on those comments either.

9.12 Dr Chapman's memorandum consisted essentially in an analysis, on a discounted cash flow basis, using both the present Treasury rate of 10 per cent per annum and a rate of 7 per cent per annum, of the comparative costs of reprocessing and extended storage. He used BNFL's documents as his basic data. Essential to this analysis was a figure for operating costs per tonne of spent fuel reprocessed and Dr Chapman arrived at a figure of £123,000 per tonne for this. Using this figure he reached the conclusion that there was no economic justification for the plant at the present time. For there to be such a justification he concluded that it would be necessary for:
i. The price of uranium to reach $60 per lb or more and
ii. Plutonium to be recycled within 10 years of recovery (in thermal reactors) and
iii. The discount rate to be 7 per cent and
iv. Overseas contracts to the full amount contemplated by BNFL to be obtained.

9.13 In addition to the matter of the figure for operating costs he made a number of assumptions including (a) that AGR fuel could be kept in cooling ponds as presently proposed for THORP for 25 years or more and (b) that THORP would be decommissioned after 10 years at a cost of £50 million.

9.14 The above items in Dr Chapman's analysis were challenged in BNFL's written comments. The figure for operating costs of THORP was said to be excessive by about a factor of five. Extended storage in ponds was considered unacceptable in the light of the evidence which I have already mentioned. It was pointed out that THORP could be expected to operate for more than 10 years. In argument it was further submitted that, if decommissioning costs for THORP were to be brought into the calculations, so also should decommissioning costs of pond storage. The criticisms in regard to extended pond storage and decommissioning costs were, I consider, valid, but they are of much less importance than the figure for operating costs and I do not consider them further.

9.15 Dr Chapman's figure for operating costs was arrived at by a process of deduction. He first took BNFL's present estimate of the price to the Japanese customers assuming that the project proceeds as planned, which price is arrived at on a cost-plus basis and is £160,000 per tonne. He than assumed a profit range of between 10 and 30 per cent and arrived at an estimated operating cost of £120,000 – £145,000 per tonne. He next took from BNFL's documents their presently estimated price to the CEGB exclusive of vitrification but 'including operating costs and return on capital employed' namely £230,000 per tonne. He deducted from this their stated capital cost per tonne of £97,000 and a further assumed £10,000 per tonne for return on capital employed. He thus reached a figure for operating costs per tonne of £123,000. Since this figure was within the range which he had deduced for the Japanese contract he used it for his analysis.

9.16 In answer BNFL pointed out, quite correctly, that such a figure for operating costs per tonne produced total expenditure on operating costs of reprocessing 3,150 tonnes of fuel (the figure used by Dr Chapman) of £387.5 million over 10 years. This they compared with Mr Allday's estimate of £140 million over 10 years for reprocessing 6,000 tonnes through THORP, that is to say the total of both UK and foreign fuel. This estimate would produce a figure for total operating costs of reprocessing 3,150 tonnes over the same period of £73.5 million, or £314 million less than Dr Chapman's figure produced. If Dr Chapman's analysis was adjusted to take account of this difference his results would be found to be broadly in line with BNFL's two documents. Dr Chapman, it was said, had fallen into error because he had failed to appreciate the make-up of the prices estimated for the CEGB and the Japanese customers respectively. He had assumed, incorrectly, that the Japanese price did not include a depreciation charge and that the price to the CEGB did not include either profit or any financing charge. In truth both prices included depreciation and profit. The essential difference between the two was that the price of CEGB included a financing charge whereas the price to the Japanese did not.

9.17 On the basis of BNFL's comments it would be necessary to deduct from Dr Chapman's range for costs on the Japanese contract a further £97,000 for depreciation in order to get to operating costs, thus getting a range of £23,000 – £48,000 for such costs. This would be in line with Mr Allday's sworn evidence, for £140 million for 6,000 tonnes gives a figure of £23,300 per ton and Mr Allday specifically stated that the price to the Japanese did include depreciation. It would also be in line with the estimated price to the home boards. If, as was stated in evidence, the price to the home boards includes a financing charge which is not included in the Japanese contract and it is assumed that the profit margin falls within the same range, the make-up of the home boards price could be:

	Home boards price	£ per tonne
Operating costs	23,000	
Capital cost	97,000	
	120,000	120,000
Profit 10% and 30%	12,000	36,000
	132,000	156,000
Levy for financing	98,000	74,000

Since the difference between the price to the home boards and the price to the Japanese is £70,000, since a charge of this order appears reasonable for its stated purpose, and since Dr Avery of BNFL stated that operating costs were only a small fraction of the total price, I conclude that BNFL's criticism of Dr Chapman's analysis is justified and that their case for present purposes is sufficiently established. I should stress, however, that it is as yet too early to reach any final conclusions on the economic position. If the project proceeds, there may be changes which will affect the position. Developments in the design, alterations in requirements for the control of emissions or a failure to obtain the amount of foreign business presently expected might each change the situation. Mr Allday, for example, made it clear that BNFL would not proceed if, in the event, the Company was unable to obtain sufficient foreign business to make the project worth while.

9.18 Before leaving the financial side it is necessary to deal specifically with the aspect of foreign business. The proposed Japanese contract was not put in evidence, for BNFL regarded it as confidential. It appeared to me that there was some danger that, if the details were made public, it might prejudice the final signature of the contract and perhaps the obtaining of further business. The contract was therefore examined by leading Counsel for FOE and WA. A summary of the principal effects of its terms was subsequently agreed and put in evidence. Having seen the proposed contract FOE's Counsel expressed the view that it appeared to be very profitable for BNFL. The summary bears this out.

9.19 Not only would it be profitable for BNFL it would be of benefit financially to the consumers of electricity in this country. The price which BNFL estimated that they would have to charge to the home boards, if a smaller plant, sufficient for UK arisings and existing foreign commitments only, was built, would be £55,000 per tonne more than if THORP were built on the proposed basis that half the capital cost would be borne by foreign customers. Additional benefits would be the earnings from reprocessing which would contribute to the balance of payments and the fact that, at the end of 10 years, BNFL would have a plant at once capable of dealing with additional home arising should this be required. The balance of payments benefit should not however be overestimated. The earnings would be spread over 10 years and might be coming in at a time when foreign earnings were not needed as badly as they are at present.

9.20 I finally mention Dr Sweet for as previously stated he was one of the principal witnesses on this subject. I need however say no more about his evidence than that I found it unconvincing.

10. Routine Discharges — Risks

System of protection

10.1 There is, nationally, and internationally, an elaborate structure to protect the public and the environment from harm from radiation. This structure was examined in considerable detail in Chapter V of the Sixth Report. The Royal Commission's principal conclusions and recommendations on the subject appear at paragraphs 527–530 and 533 of that Report. The Government's response to those recommendations are to be found in paragraphs 13–31 of and in Annex A to the White Paper. It would serve no useful purpose if I were to repeat the examination in this Report. The essential question which I have to consider is whether the system is such that, if outline planning permission for THORP is given, the system can be relied upon to give adequate protection from harm to workers, public, future generations and the environment. If it can, then it is the task of that system to afford the necessary protection and not that of the planning authority. There are basically three ways in which it could be shown that the system could not be relied upon. These are
 i. If the operation of THORP would necessarily involve the release of radiation at higher than tolerable levels.
 ii. If the system itself was defective.
 iii. If the competence capability or integrity of the bodies which together make up the system was in doubt.
I shall first describe the system in broad outline and then examine each of these questions.

10.2 The principal international body in the system is the International Commission on Radiological Protection (ICRP). This body recommends, but has no power to fix, radiation limits, adherence to which it considers will sufficiently protect man from harm. It is independent of any government and its members are selected on the basis of their scientific reputation and standing. ICRP does not recommend limits designed to protect the environment generally. It considers that if man is sufficiently protected, so also will be vegetation, birds, beasts etc.

10.3 Other international bodies concerned with the protection of man from harm from radiation but which have, so far as the United Kingdom is concerned, no more than advisory powers are (1) the United Nations Scientific Committee on the Effects of Atomic Radiation (UNSCEAR), which meets annually and reports on levels of radiation from different sources and on the scientific evidence of their effects (2) the IAEA, mentioned already in connection with non-proliferation (3) the World Health Organisation (WHO) (4) the Food and Agriculture Organisation (FAO) (5) the Nuclear Energy Agency (NEA) of the Organisation for Economic Co-operation and Development (OECD) which, through one of its committees which meets twice a year, maintains a continuous review of radiation protection standards and is, in conjunction with IAEA, concerned with radioactive waste disposal in the deep ocean.

10.4 A further international body, also mentioned previously in connection with non-proliferation, is EURATOM. I mention this body separately because, unlike those previously mentioned, it does have power to, and does, fix standards which bind the United Kingdom. Such standards are however based on ICRP recommendations so that, effectively, the international limits which apply to the UK are ICRP limits.

10.5 The foregoing shows clearly that basic recommendations are arrived at by a highly qualified independent body and that standards are kept continuously under review internationally.

10.6 ICRP recommended limits and EURATOM binding limits are, however, upper limits only. The United Kingdom is therefore free to fix such lower limits as it sees fit. I describe in the next following paragraphs the structure as it will be as a result of the Government response to the Royal Commission's recommendations.

10.7 In accordance with the Royal Commission's recommendation (para 533 conclusion 40), responsibility for nuclear waste management policy will lie with the Secretary of State for the Environment together with the Secretaries of State for Scotland and Wales instead of, as hitherto, being divided amongst a number of departments (White Paper para 14). A Nuclear Waste Management Advisory Committee will also be established for the purpose of advising the Government on waste management policy. It may be given further functions, such as the initiation of long-term research, but this is as yet undecided. (White Paper paras 16 and 17).

10.8 The NRPB will advise the Government, by virtue of a direction under the Radiological Protection Act 1970, on the adequacy for the UK of ICRP/EURATOM standards. The Medical Research Council (MRC) will provide advice as to the biological bases upon which such standards rest (White Paper para 23).

10.9 The position with regard to the fixing of standards thereafter is more complex. Under Section 6 of the Radioactive Substances Act 1960, discharges are permitted only in pursuance of a joint authorisation of the Secretary of State for the Environment and the Minister of Agriculture, Fisheries and Food. Their respective responsibilities are in practice exercised, in relation to discharges to the atmosphere, by HM Alkali and Clean Air Inspectorate (ACAI) and, in relation to discharges to water, by the Fisheries Radiological Laboratory (FRL). In arriving at the limits of the discharges which they authorise both bodies act upon three principles set out in a White Paper of 1959 entitled 'The Control of Radio Active Waste' (Cmnd 884 1959) (BNFL 83). This White Paper is currently under review. The three principles are

'a. to ensure, irrespective of cost, that no member of the public shall receive more than (the relevant ICRP dose limits for the whole body);

b. to ensure, irrespective of cost, that the whole population of the country shall not receive an average of more than 1 rem per person in 30 years and

c. to do what is reasonably practicable, having regard to cost, convenience and the national importance of the subject, to reduce doses far below these levels'. (Cmnd 884 para 117)

10.10 The first and third of these principles closely correspond with two of the main features of ICRP's recommended dose limitation system namely

i. 'the dose equivalent to individuals shall not exceed the limits recommended for the appropriate circumstances by the Commission' and

ii. 'all exposures should be kept as low as reasonably achievable, economic and social factors being taken into account'. (ICRP 26 para 12) (G 35)

10.11 The second principle applied a limit which was one-fifth of ICRP's then recommended limit for population dose. The Commission have, in ICRP 26 their most recent publication, abandoned any recommendation with regard to population dose because they considered that the former limit was unlikely ever to be reached and might suggest the acceptability of a higher level of dose than was necessary and a higher risk than was justified. They also considered that genetic effects, for the restriction of which the population dose limit had been recommended, were, on evidence accumulated over the past two decades, unlikely to be of overriding effect. If their current limitations on individual dose limits were observed the average dose to populations would in their view be well within acceptable levels. (ICRP 26 paras 129 and 130).

10.12 It is also necessary to mention disposal of solid waste and the accumulation of waste in whatever form. Both also require authorisations from the Department of the Environment (DOE) and/or the Ministry of Agriculture, Fisheries and Food (MAFF) under Sections 6 and 7 respectively of the Radioactive Substances Act 1960.

10.13 In Scotland all authorisations are issued by HM Industrial Pollution Inspectorate on behalf of the Scottish Development Department after consultation with the Department of Agriculture and Fisheries for Scotland. This latter department itself authorises the sending of solid waste to sea for ocean dumping.

10.14 The responsibility for determining and controlling discharges involves, if it is to be so exercised as to afford proper protection to the public, both extensive research and extensive monitoring of results of discharges. In order to determine levels of discharge the controlling authority needs to know the ways in which, and the levels at which, released radioactivity can get back to man or the environment and the effects which it will have when it does so.

10.15 Extensive research has been carried out by a variety of bodies in the UK of which the principal ones are MRC, NRPB, FRL and the UKAEA. Equally extensive research has also been carried out by international bodies such as I have already mentioned and by national bodies in other countries. To such research must be added that done by universities and individuals all over the world. The volume of published material can only be described as enormous.

10.16 In certain areas the benefit of world-wide research accrues to the United Kingdom authorities. For example, work on the effects of a particular radionuclide, when ingested or inhaled by man, is of general application. In other areas the necessary research for the protection of the public can only be done in the United Kingdom and is, to a large extent, only applicable to the United Kingdom. What happens to discharges to the Irish Sea from Windscale is a simple example. The radionuclides will be dispersed in the sea but they must then be followed. It must be ascertained, for example, to what extent they are taken up by various kinds of fish, which may be eaten or turned into fertilisers, or by seaweed, which may also be incorporated into foodstuffs or fertilisers, whether they are redeposited on land and if so where, or whether they get or can get back to man in the form of sea spray. But such matters are only the beginning of the research operation. Next it must be ascertained how much of the radioactivity in the fish or seaweed or deposited on the shore gets back to the most exposed members of the public. Extensive research of this kind has been done, principally by FRL.

10.17 Extensive as the research has been no-one suggested or could suggest that all the answers are known. Following upon the announcement in the White Paper of the new responsibilities of the Secretary of State for the Environment in relation to nuclear waste management, NRPB submitted a list of studies they considered necessary for assessing the environmental effects of discharges. A copy of this list was produced in evidence at my request. It contains some 22 suggestions. This does not necessarily mean that research hitherto has been inadequate, for a main object of research is to keep ahead of problems, and what may be necessary at one stage may have been wholly unnecessary at an earlier stage. The Royal Commission made a number of recommendations with regard to improvements in research and monitoring (Sixth Report paragraphs 528–529). These were not wholly accepted in the White Paper. The system which will now be in operation is as follows:

i. The Secretary of State for the Environment together with the Secretaries of State for Scotland and Wales will ensure that there is adequate research and development on methods of waste disposal (White Paper paragraph 25).
ii. DOE will review, in conjunction with other bodies concerned, the adequacy of the present research programme. This review will include the question of research into the effects of radiation upon the natural environment which the Royal Commisssion regarded as having been insufficient (White Paper Annex A paragraph 12).
iii. There will be a joint committee of NRPB and MRC to co-ordinate research.
iv. FRL will continue its monitoring activities.
v. Atmospheric discharges will be monitored by ACAI.

Risk levels if THORP is built

10.18 I now turn to the question raised in paragraph 10.1 (i) above, namely whether operation of THORP would necessarily involve exposure of the public to higher levels of radiation than are tolerable. This question appears at first sight to be of crucial importance for, if the answer were to be in the affirmative, my recommendation would clearly be that permission should be refused. I have however no hesitation in saying that it is a question which neither I nor anyone else can answer now, either in the negative or in the affirmative. If permission were given promptly and acted upon equally promptly, THORP would not begin to operate for 10 years. BNFL have requested that if permission is granted they should have seven years in which to act upon it before it lapses. It may therefore be that the period before THORP would be ready to operate would, quite apart from the possibility of over-runs in building time, be even longer. It was repeatedly stressed to me by objector after objector that, in the nuclear field, developments in technology, in radiobiological knowledge, in public attitudes and in policy, change with great rapidity. It did not need stressing. The situation is very obvious. Within the next few years much may happen. Present limits may be reduced so that what appears safe now will be accepted as unsafe. Or they may be relaxed and what appears unsafe now may be accepted as safe. Technological developments may result in it being possible to contain releases well below what is now believed possible, or to burn up highly active waste and thus avoid the need for glassification and subsequent disposal. Risks presently considered by some to be intolerable may become accepted as tolerable e.g. because increasing knowledge of the effects of coal burning may show that that alternative is very much worse; or risks presently accepted as tolerable may come to be considered intolerable e.g. because alternative sources of energy or conservation measures may make it unnecessary to tolerate any pollutant source of energy at all, be it from nuclear or fossil fuel stations.

10.19 In such a situation no final answer to the question is possible and it was argued that, because of this, planning permission should be refused. It is convenient to give one example of the way it was put.

10.20 Dr Alice Stewart, a witness for the Town and Country Planning Association (TCPA) to whose evidence I shall return later, had conducted research jointly with two others. As a result, she and her co-workers considered that the cancer risk from low level radiation might be up to 20 times greater than the currently accepted estimate. She then stated:—

'it should be clear that the evidence relating to the safety of the public and other aspects of the national interest . . . is disputed among experts, and consequently . . . no decisions on the proposals before the Inquiry should be made until the issues have been satisfactorily resolved'.

10.21 It is probably the case that, within two or three years, the particular matters advanced by Dr Stewart will have been resolved and that her conclusions will either have been accepted, or proved to be, in part or in whole, ill-founded. But that will not resolve the matter for there will almost certainly be other issues on which there is dispute among experts. Indeed it is to be hoped that there will, for it is by the challenging of generally accepted opinion that it is put to the test and that the public protection is best assured.

10.22 If therefore a substantial dispute amongst experts is a good ground for delaying a decision it follows either:
a. that there would never be a decision at all because at the time of consideration there would be an existing dispute, or
b. that there could only be a decision if, at the time at which the decision fell to be made, there happened to be a temporary pause in the stream of scientific criticism.

10.23 Both results appear to me equally unacceptable. The argument for delay on this ground fails to take into account the nature of the application. The proper bodies to evaluate Dr Stewart's conclusions, and a number of other matters raised before me, are such bodies as, internationally, ICRP, UNSCEAR and NEA and, nationally, NRPB, MRC and UKAEA. The proper bodies to take any action which may be necessary as a result of such evaluation are the control authorities.

10.24 That this is and must be so is simply illustrated. Outline planning permission has already been given for extensions to the magnox reprocessing plant. The intended discharges from the magnox plant after the extensions have been completed will be less than they are presently and less, in certain respects, than the intended discharges from THORP. If, however, dose limits were, for whatever reason, reduced to a point at which the intended discharges would exceed the new limits it would be no answer for BNFL to say 'Oh but we have planning permission'. That fact would not even be relevant. BNFL would have to comply with new limits.

10.25 Notwithstanding the views expressed above it would however be quite wrong for me to leave the question under consideration with no further examination. This is so for a variety of reasons, of which the principal ones are, first, that if, on the evidence before me, I felt that there was a real likelihood that THORP could not operate at tolerable levels I should feel it not only right to say so but my duty to say so and, secondly, that there are a number of matters of importance arising on the evidence upon which I was invited to express my views, whether or not they were strictly planning matters. In view of (1) the large amount of evidence tendered on these matters, much of it by eminent persons from this country and overseas, and (2) the time, trouble and expense involved in tendering that evidence and making submissions upon it, I have felt it right to accept that invitation.

10.26 The current assumption in radiological protection practice is that any additional amount of radiation, however small, may do some harm. One party advanced the view that there was a threshold dose below which no risk of harm at all would occur, but this is against the weight of the evidence and I do not accept it. Since it was not suggested that THORP could be built without exposing workers and public to some degree of additional radiation, however small, it follows that, if no amount of radiation is tolerable, THORP cannot operate at tolerable levels of discharge. This extreme view was advanced. I reject it. If it were right then, among other things, all fossil fuelled plants would have to be closed down (see para 8.40 to 8.42 above).

10.27 The questions which next arise are: (a) what is the amount of harm or risk of harm likely to result from THORP? and (b) is such harm or risk of harm tolerable? It was generally accepted that it was for the scientist to assess the former but there was much argument about (1) the method of assessment of the latter, (2) the level of harm or risk which should be regarded as tolerable and (3) the manner in which the level of risk, whatever it might be, should be explained to the public. These matters may be thought more appropriate for consideration when I come to deal with the adequacy of the system of protection but I find it impossible to consider the likelihood of THORP being able to operate to tolerable levels without dealing with them at this point.

10.28 I take first the question of the explanation of risk. Professor Fremlin, Professor of Applied Radioactivity at the University of Birmingham, and a member of the Campaign for Nuclear Disarmament, who gave evidence for Cumbria in support of BNFL's application, used as a yardstick for comparison with radiation risks, their equivalents in number of cigarettes smoked per day or per week. For so doing he was severely, indeed bitterly, criticised by other witnesses. In so doing he was by no means alone. The Royal Commission, for example, observed in paragraph 52 of the Sixth Report that radiation workers who received an annual radiation dose of 1 rem were running a risk of about 1 in 10,000 that they would eventually die of cancer as a result of each year's dose and that was approximately as dangerous as regularly smoking three cigarettes per week. The criticism was levelled because the risk from radiation is, to a large degree, an unavoidable risk. It was therefore considered wrong to compare it with the voluntary risk of smoking. This criticism is not in my view valid. If a decision making body is trying to assess whether the public will accept a particular risk, it is, in effect, asking itself the question 'are the public likely voluntarily to accept this risk?' I see therefore no objection to looking at other voluntarily accepted risks.

10.29 In Table 8 and paragraphs 170 and 171 of the Sixth Report the Royal Commission broadened its base of comparison to include a variety of risks some wholly voluntary, others partly voluntary and partly imposed, and yet others wholly imposed. I have no doubt that the best way to explain the degree of risk to the public is to give a broad range of comparables. Different individuals will find some comparisons of more use than others. If a particular risk is stated to be 10^{-6} (the way experts describe risks), or 1 in a million, or as being a 100 per cent increase on a naturally existing risk, it means little or nothing to the ordinary person. If, however, such a person is told that, as has been estimated, such risk (1 in a million) is the same as that involved in smoking $1\frac{1}{2}$ cigarettes, travelling 50 miles by car or 250 miles by air, rock climbing for 90 seconds, canoeing for 6 minutes, engaging in ordinary factory work for 1–2 weeks or simply being a male aged 60 for 20 minutes, it would mean a great deal to him.

10.30 If a man is asked if he will accept a 100 per cent increase on a natural risk he may well be alarmed and say that he will not. If he is told that the increased risk is merely the same as any one of the above list, particularly perhaps the last, he would probably consider himself, and be considered by others, to be of an exceptionally timorous nature if he declined; the more particularly if declining involved depriving others of benefits, as for example the ability to obtain work. I am reinforced in this view by the fact that Dr Wynne who represented the NNC accepted that there was no better way to explain risks to the public. I shall use this method to explain what appear to me to be the risks from routine discharges from the proposed plant.

10.31 Current ICRP limits for whole body doses are 5 rem per annum for radiation workers and 500 mrem per annum for individual members of the public. There are also basic limits for body organs, if exposed alone, which, in general, are higher than the corresponding dose limits for whole body exposure. These limits are the basic standards. In addition there are a large number of secondary limits of general application which are designed to ensure that the basic standards are not exceeded. Examples of such limits are maximum permissible concentrations in air or water for the various radionuclides (MPC_a and MPC_w), maximum permissible body burdens (MPBB), maximum permissible lung burdens (MPLB) and so on. Additionally there are specific derived working limits for discharges (DWLs) which are designed to ensure that if the discharge limit is not exceeded no-one shall receive a dose greater than the basic standard.

10.32 On current risk estimates a radiation worker receiving the full dose limit of 5 rem annually incurs the risk of about 1 in 2000 that he will *eventually* die of radiation induced cancer from each year's exposure (Sixth Report para 52). This is approximately the same as the risk run by every member of the public that he will die of some form of accident in *that* year. A member of the public receiving the full dose limit for the public will incur about 1/10th of that risk. He will thus be 10 times more likely to die from an accident of some sort in the year in which he receives the dose than he will be to die of cancer years later as a result of that dose. Neither risk appears to be particularly severe.

10.33 BNFL's intentions are, however, that maximum doses to workers and public alike shall be very much lower than the basic limits. They are that the dose to workers shall be limited to 1 rem per annum and that the dose to individual members of the public from total discharges from all reprocessing operations, both THORP and magnox, shall be kept to about 50 mrem per annum. The dose to workers of 1 rem per annum is the equivalent of regularly smoking three cigarettes a week and one-fifth of the risk of accidental death in the year in question. BNFL's intentions with regard to the public involve a risk 20 times lower than this. A member of the public receiving the intended maximum dose will therefore be 100 times more likely to die in an accident of some sort in the year in which he receives it than to die years later from cancer as a result of it. He will be about 10 times more likely to die of leukaemia from natural causes and, if he regularly smokes 10 cigarettes a day, he will be 500 times more likely to die from that cause than from emissions from THORP and magnox combined. The annual risk from this source will be about the same as that involved in travelling 250 miles by car or being a male aged 60 for 1 hour 40 minutes.

10.34 If current estimates are correct, and if BNFL fulfils its intentions, it seems to me impossible to suggest that any substantial numbers of the public or of workers would regard the risks as intolerable. As to workers, Mr Adams, National Officer of the Electrical, Electronic, Telecommunications and Plumbing Union, and Chairman of the Trades Union side of BNFL's Joint Industrial Council, gave evidence before me and was firmly in favour of the proposals. It was clear that he was speaking with a considerable knowledge of the subject, in particular of BNFL's previous record. As to the public, I find it difficult to believe that there are many, perhaps any, so lacking in generosity, that they would refuse to accept the risk involved if to accept it would provide some demonstrable benefit even if that benefit were small. Indeed even if BFNL's intentions failed to such an extent that members of the public received the full permitted limit I would have a similar difficulty. It is possible, I suppose, that there are people who would say that although they were, each year, 10 times more likely to die of some accident, they found such a risk intolerable and would rather the benefit were denied than accept it. I do not believe there can be many such people. Accordingly, if current estimates are right and BNFL's intentions are fulfilled, or even if a combination of errors in the estimates and a failure in BNFL's intentions resulted in the risk being 10 times higher than I have assumed, my own opinion would be that the risk would be tolerable, so far as workers and individual members of the public are concerned.

10.35 The collective results from population exposure both somatic and genetic have also to be considered. With regard to the former it is presently estimated that a collective dose of 1 million man-rem would lead to 100 fatal results, of which about $\frac{1}{4}$ might be leukaemias (Sixth Report para 52). BNFL produced estimates of the collective dose commitments if THORP were to operate to full capacity for ten years. These show on the basis of the above estimate that the total number of cancers likely to result from such operations would be between 2 and $2\frac{1}{2}$ per year, or about 1 per year if krypton 85 were to be efficiently removed from the atmospheric discharges. (See Annex 3). This degree of harm can usefully be compared with the harm which would be likely to result if there were no reprocessing. In that event reactors would need to be fuelled entirely with mined uranium. The requirement of mined uranium to produce an in-put of fresh fuel to reactors and an out-put of spent fuel to THORP of 1,200 tonnes is about 5,000 tonnes. Use in thermal reactors of the uranium and plutonium recovered from reprocessing the out-put of 1,200 tonnes would provide about 35 per cent of the reload requirement i.e. the equivalent of 1,750 tonnes. The number of deaths to uranium miners likely to result from mining this quantity is between 3 and 4 per annum. The saving in deaths in uranium mining would of course be very much greater if the plutonium were used in FBRs but I do not use this for comparison. There may never be FBRs. On the other hand it seems almost inevitable that both the uranium and plutonium recovered from reprocessing will be used in thermal reactors if there are never any FBRs. I observe also that if the energy which could be derived from the recovered uranium and plutonium were to be provided instead by coal burning stations the resulting deaths would be greater. Bearing in mind these alternatives I regard the degree of harm as plainly tolerable. Indeed I find it difficult to understand the process of thought which appears to find it preferable to avoid a death from radiation at the expense of causing more deaths from other causes.

10.36 The total number of substantial genetic abnormalities which would be induced by radiation in all subsequent generations was estimated (by UNSCEAR and by the BEIR Committee) in 1972 as 300, and (by UNSCEAR and by ICRP) in 1977 as 200, per million man-rem. The estimated releases from THORP would involve genetically significant radiation levels which would cause about one such case (on the higher estimates, which were those quoted in the Sixth Report) resulting from each year of operation. This figure would be reduced by about 25 per cent if krypton 85 were not released. As with the estimated frequencies of cancer induction given in para 10.35, these figures include the effects of the occupational radiation exposure as well as those in the general public (see Annex 4).

10.37 The next matter for consideration is whether BNFL are likely to achieve their intentions. I have no hesitation in concluding that there is every likelihood that they will do so and that the possibility of their failing to do so to such an extent as to reach the ICRP dose limits is very remote. I had the advantage of seeing and hearing BNFL's witnesses under cross-examination and I share the confidence expressed by Mr Taylor on behalf of PERG that their designer Mr Warner could design to meet any standard which was set. I have equal confidence that Mr Warner's designs could and would be put into effect. I do not mean by this that there will be no occasions when things go wrong. Errors of one sort or another occur in any human operation and will beyond all doubt be made by BNFL. They have made errors in the past and will do so again. I shall give two examples of recent errors which involved substantial releases of radioactivity beyond the perimeter of the site. In 1972 the average three-monthly discharge of iodine 131 from Windscale rose suddenly from 0.14 curies to 5.5 curies, a 39-fold increase. This was due to the inadvertent reprocessing of fuel which had not been cooled for the proper time. The following year the average three-monthly discharge had reverted to 0.30 curies. This was clearly a serious error but it must be seen in perspective. The effect of an additional discharge of this size to members of the public would be small.

10.38 The second example relates to discharges of caesium to the sea. In 1970 the discharge of caesium 137 rose from 12,060 curies in the previous year to 31,170 curies. It stayed at about that level for the next two years, dropped to 20,770 curies in 1973 and then rose to 109,770 curies in 1974 and to 141,377 curies in 1975. The last figure available is that for 1976 when the discharge was 115,926. For simplicity I omit the corresponding figures for caesium 134. The cause of this increase was the corrosion of magnox fuel cladding in the cooling ponds and a consequent leakage of caesium into the pond water. The reduction in 1976 was due to the introduction, in my view belatedly, of measures to reduce the discharge. Such measures are however limited in their effectiveness and the situation will only markedly improve when the pond water treatment plant, which is included amongst the magnox extensions for which permission has already been granted, comes into operation. This is expected to be in 1980–81. When it does come into operation it will, if THORP is built, also treat pond water from the THORP ponds. A new authorisation is currently under discussion which if implemented would limit the total discharge of caesium to a maximum of 40,000 curies per annum, about three times lower than the present rate. Despite the very great increases mentioned, however, FRL's calculations of doses to fish eaters in 1976 (which were, owing to retention times of caesium in the Irish Sea, higher than in 1975 notwithstanding the lower discharge) only reached the levels set out in Table 7 to Part I of their Annual Report for 1976 (BNFL 160) which I reproduce here for convenience.

Maximum rates of radiation exposure from Windscale discharges in 1976 due to consumption of fish and shellfish from the Irish Sea

Population group and persons concerned	Assumed consumption rate and source	Radiation exposure (% of ICRP-recommended dose limit) Total body
Coastal fishing community: maximum consumer	265 g/day Local supplies	44
Coastal fishing community: average consumer	52 g/day Local supplies	9
Other fish-eaters: Critical group average	300 g/day Commercial: Whitehaven/ Fleetwood landings	17
Public at large: typical consumer	40 g/day Commercial: Whitehaven/ Fleetwood landings	2.4

10.39 The foregoing examples demonstrate that there is a large margin of safety and that even when serious errors are made the results need not endanger the public. Indeed if the system works properly and the limits are right there will never be such danger, for if the dose limits were being approached, the control authorities would step in. In this connection it is as well to mention that the dose limits are not intended to set some absolute standard, the exceeding of which will at once result in a dangerous situation. Dr V T Bowen, Senior Scientist of the Woods Hole Oceanographic Institute, USA and a witness for IOM, expressed the view that the situation (assuming the limits were right) would not become dangerous until the limit had been exceeded by 100 per cent or more.

10.40 I limit mention of BNFL's past errors to the two foregoing examples for, although their past record was subjected to very close scrutiny as a basis for submissions that they could not be relied on to fulfil their intentions sufficiently for the protection of the public, I am satisfied that, with the large margin of safety which exists, the submission is ill-founded.

Risk levels—Suggested inadequacies of current estimates and limits

10.41 Having concluded that BNFL are likely to achieve their intentions I now consider whether there is a real likelihood that risks currently estimated are so far wrong that THORP could not be built and operated at tolerable levels. To cover every suggestion advanced concerning alleged defects in the limits would be inappropriate. I shall take in turn only those suggestions principally relied on by objectors and consider them as briefly as possible.

Dr Alice Stewart

10.42 Dr Stewart and her co-workers concluded that cancer risks might have been under-estimated as much as 20 times. This conclusion was reached on the basis of a paper (the Mancuso, Stewart and Kneale paper) (IOM66) which had not been upblished at the conclusion of the Inquiry but which has since been published (Health Physics *33*pp. 369–385 1977). It was, however, known about in scientific circles by December 1976, and had attracted criticism from a number of sources. It was based on data relating to workers at the American nuclear establishment at Hanford. Dr Stewart herself and her co-worker Mr G Kneale both gave evidence before me. There can be little doubt that if Dr Stewart's conclusion is valid it would seriously affect the whole picture. This was expressly accepted by Dr Dolphin of NRPB. It would not however necessarily mean that THORP could not be built to tolerable levels. If the permitted dose were reduced to 1/20th of its present level it might still be possible to build and operate the plant to comply with that level. If it proved to be impossible then it would have to be abandoned.

10.43 I have mentioned that the Mancuso, Stewart and Kneale paper had met with criticism from various sources. As a result Dr Stewart had already made a number of amendments to the original paper prior to giving evidence at the Inquiry. One source of such criticism was Professor J Rotblat, Emeritus Professor of Physics at the University of London, who, like Dr Stewart, was called by TCPA. At the public hearings on the projected CFR-1 at the London International Press Centre on 13 and 14 December 1976, Professor Rotblat stated that he did not accept 'the report which came out a few days ago that, in the United States, (radiation) workers have had an increase in the incidence of cancer'. He thought that the samples were too small at that stage to enable an opinion to be expressed one way or the other. The report to which he referred was either the original or an early version of the Mancuso, Stewart and Kneale paper. When asked about this matter in evidence he stated (i) that before the public hearings in December he had only had an opportunity to glance at the paper and could not quite understand it, (ii) that he had since discussed the paper with Dr Stewart on several occasions to try to understand it, (iii) that Dr Stewart had on such occasions been very convincing but he still did not understand it fully and (iv) that he still did not accept the results, albeit his non-acceptance was less emphatic. Such a view, coming from such a person, appears to me of considerable significance. Dr Stewart was convincing in evidence in the sense that she rejected with supreme confidence suggestions that her results were wrong but she failed to deal with a number of what appeared to me to be valid criticisms. It is right that having made such a statement I should give instances. I shall give three.

10.44 *Example 1*
  a. In Table 2 of the paper as presented at the Inquiry Dr Stewart set out data relating to 3520 Hanford workers who had died. This group was divided into those who had died of cancer and those who had died of some other cause. The table recorded that workers who had been exposed to radiation accounted for 66 per cent of the cancer deaths but only 61.1 per cent of non-cancer deaths and that this difference was statistically significant. Dr Stewart explained in evidence that this result, which was noted at an initial stage in the research, was such as to indicate that there might be something happening—i.e. some connection between radiation exposure and cancer deaths, and that it led her and her co-workers to go further.
  b. In April 1977 Dr Ethel S Gilbert of the Battelle Memorial Institute USA commented upon the paper in its revised edition as at March 1977 (BNFL 311). Amongst her comments she included comment on Table 2. This comment was that if the exposed workers, who had been taken by Dr Stewart as one entire group, were divided into two groups, those employed for less than two years and those employed for two years or more, the apparent difference in Table 2 was eliminated. Thus

	% occurring in exposed workers	
	Cancers	Non-cancers
All deaths (from Table 2)	66.0	61.1
Deaths among those employed < 2 years	33.9	33.9
Deaths among those employed 2+ years	87.4	87.6

  c. Dr Stewart was aware of this comment in April 1977. She did not, however, mention it or seek to refute it in her evidence in chief although she both mentioned, produced and sought to deal with an earlier (July 1976) report by Dr Gilbert on the Hanford data in which the conclusion had been reached that 'analysis of the full data does not exhibit any evidence of a relationship of radiation exposure and cancer as a cause of death'.

d. When Dr Gilbert's comment on Table 2 was put to Dr Stewart in cross-examination her observation was merely that lines two and three of the table at b. above were totally incompatible with the first line 'because an overall figure of 66 per cent could not arrive at a sub-division of 33.9 and 87.4 when a 61.1 gives you 33.9 and 87.6. I think that Miss Gilbert probably made a little arithmetical error'. In Dr Stewart's view the figure could not possibly be correct.

e. Dr Stewart's observation was made with complete confidence and was thereby on the face of it 'convincing'. Indeed it might be thought somewhat patronising towards Dr Gilbert. It was, however, clearly wrong. The figures are plainly algebraically *possible* and the apparently significant difference in the overall rates (66 per cent and 61.1 per cent) could be accounted for and shown not to be significant if (i) there was a higher percentage of cancers in the 2+ year group (both exposed and non-exposed) than in the < 2 year group and (ii) the percentage of exposed workers in the 2+ year group was higher than in the < 2 year group.

f. Although I was able to satisfy myself algebraically that Dr Stewart's answer was untenable I had not the basic Hanford data available to me. I therefore asked that I be provided with a revised version of Table 2 dividing the exposed workers into the same two groups as Dr Gilbert had used and giving the basic data.

g. As a result I was provided on the last day of the Inquiry with two further tables prepared by Dr Stewart. The first was a revised Table 2 sub-divided on the basis which I had requested. This contained figures which, although not exactly the same as Dr Gilbert's, confirmed that the division into the two groups eliminated any significant difference between the percentages of deaths from cancers and non-cancers occurring in exposed workers.

The second table consisted in a further revision of Table 2 in which causes of death, instead of being divided merely into cancers and non-cancers, were divided into RES Neoplasms, other cancers, accidents and other causes of death. This Table showed on its face that in both the 2+ year group and the <2 year group those dying of RES neoplasms had the highest radiation doses. Surprisingly, however, whilst the percentage of deaths from such neoplasms occurring in those who had been employed, and thus exposed, for the shorter period (<2 years) was shown to be statistically significant this was not the case in those who had been employed, and thus exposed, for longer periods (2+ years).

h. I was unable to investigate whether the last Table had any greater significance than the earlier one but the somewhat surprising result mentioned above coupled with the manner in which Dr Stewart had dealt with Dr Gilbert's criticism gave me no confidence that it had.

10.45 *Example 2*

(a) *Table 11* of the paper was designed to test for a correlation between the percentage of cancer deaths and the cumulative radiation dose after standardisation for age at death. The final figure in the Table was 0.46 ± 0.22 and this was regarded by Dr Stewart as sufficiently high to be significant in establishing a correlation between cancer deaths and cumulative radiation dose.

(b) The Table was divided, as its title suggests, into age groups, one of which was the group aged from 60–69. There were 239 deaths from cancer in this group. The group itself, like all other groups, was divided into five sub-groups according to the radiation doses received. It was demonstrated in cross-examination of Mr Kneale that if one of those dying of cancer in the sub-group with an accumulated dose of 500+ centirads, had reached his 70th birthday, before dying, thus reducing the number in that group by 1 (from 12 to 11) and increasing the number in the next group by 1 (from 5 to 6) the final figure would have been 0.32 instead of 0.46. Mr Kneale agreed that such lower figure would not be significant in establishing a correlation.

(c) It was later established that a different move of a single person from one group to another could also increase the figure of 0.46 and thus show an apparently more significant correlation.

(d) It may be that in reaching such a conclusion I am flying in the face of statistical theory but my own conclusion is that if the significance or otherwise of an apparent result can depend on the chance that a single man died just before rather than just after a particular birthday, the result shown is not convincing.

10.46 *Example 3*

(a) Table 4 of the paper was entitled 'Observed and Expected Numbers of Specific Neoplasms listed according to Mean Cumulative Radiation'. It listed 18 types of neoplasms and used for expected deaths the figures from NCI Monograph 33 for cancer deaths for White US males in 1960.

(b) Dr Stewart, when cross-examined about this, at first asserted that the year 1960 had been carefully chosen because over 50 per cent of he deaths were before 1960; but she later agreed that there were a substantially higher number of deaths after 1960 than before. She also agreed that cancer rates in the United States had been increasing since 1960.

(c) I am unable to attribute much, if any, value to figures which do not correlate observed deaths with deaths expected at the dates when the observed deaths occurred, the more particularly when the attempt to justify the 1960 comparison as carefully chosen as a 50/50 figure was swiftly acknowledged to be wrong.

10.47 Although I found Dr Stewart's interpretation of the Hanford data (the present paucity of which in certain respects was acknowledged by Mr Kneale) unconvincing, I should perhaps stress that I have no doubt about either the importance of such data or the desirability of accumulating data of the same nature about radiation workers in the United Kingdom, a matter which was referred to in paragraphs 74 and 75 of the Sixth Report. Arrangements for such accumulation are already in hand. I therefore say no more on the subject except that I have not relied upon the paper by Dr Dolphin of NRPB (NRPB R54)–(BNFL 199) reviewing figures relating to Windscale workers in reaching my conclusions on Dr Stewart's evidence or indeed on any other matter.

Professor Edward Radford

10.48 Professor Radford, a member of US National Academy of Sciences Advisory Committee on the Biological Effects of Ionising Radiation and of that Academy's present committee which is engaged in up-dating the Report of the earlier committee (the BEIR report), was called on behalf of NNC. He was not an opponent of nuclear power nor did he advocate a 'nil' release of radioactivity in the course of operating nuclear establishments. He did however consider that the present risks of cancer, in particular the lung cancer risks, were underestimated and that the MPC_a for insoluble plutonium and americium should be reduced by a factor of 200.

His recommendations may be summarised as follows:—
  i. The whole body dose limit should be reduced to 25 mrems p.a., i.e. to 5 per cent of the present ICRP limit.
  ii. the MPC_a for plutonium and americium should be reduced by a factor of 200.
  iii. All releases of tritium and plutonium should so far as practicable be to sea and not from the stacks.
  iv. There should be included in THORP, if built, plant for the containment of krypton 85.

10.49 If THORP were built and operated so as to comply with these recommendations he would have no objection to it. He considered, however, that before final design and construction of THORP was permitted the magnox facility should operate to his recommended limits for at least three years in order to demonstrate BNFL's ability to operate to such limits. He finally invited me to request the Secretaries of State for Energy and the Environment to call an international meeting of countries concerned with nuclear power development to resolve differences concerning policy and standards. I record the invitation, but I do not accept it. Any recommendation for such a meeting could only be made if there was no adequate international machinery for fixing standards. I consider that there is such machinery.

10.50 I am satisfied that there should be, and that there will be, further research, both into cancer risks generally and into appropriate secondary limits for particular radionuclides, but the material upon which Professor Radford's conclusions were based appeared to me unsatisfactory in a number of respects. He referred for example to studies in relation to uranium miners in Sweden and Czechoslovakia. The Swedish results had not yet been fully evaluated and he accepted that they were of too preliminary a nature to be of value. In the case of the Czechoslovak miners the studies did not disclose the numbers of exposed miners. He also referred to studies relating to patients suffering from ankylosing spondylitis who had been given radiotherapy but there are as yet no valid estimates of the doses to the lungs and bronchi of such patients.

10.51 I do not therefore consider that his conclusions have sufficient weight to justify me in finding that the present limits are likely to be changed so radically as to suggest either that THORP cannot be built to tolerable limits or, equally important, that the public are or have been under any serious risk from present or past releases.

10.52 With regard to tritium and krypton, Professor Radford regarded his suggestion as to tritium as an additional precaution only and in this context I need say no more about it, the more so as it is BNFL's intention to discharge all tritium to sea in any event or, at most, only a small fraction to atmosphere. With regard to krypton it is accepted by BNFL that krypton removal plant will be incorporated if the technology for its removal and safe retention is available. I am satisfied that it should. I also consider that BNFL should not merely stand by and instal such a plant if and when others develop it. They should themselves devote effort to its development.

Dr Sadao Ichikawa

10.53 Dr Sadao Ichikawa, Professor at the Laboratory of Genetics, Kyoto University, Japan was also called on behalf of NNC. He was wholly opposed to nuclear power and was not personally prepared to accept any amount of radiation exposure, however small, from nuclear establishments. He considered that the doubling dose for genetic effects should be 10 rem or lower. The doubling dose is the dose estimated to double the number of naturally occurring mutations. Existing estimates range from about 10 rem to 100 rem. Dr Ichikawa's figure is therefore at the lower end of the range. As the Royal Commission pointed out, less is known about the incidence of genetic effects per unit of dose than in the case of cancer induction. Dr Ichikawa based his conclusions largely on work which he had carried out on the plant Tradescantia but accepted that a similarity between Tradescantia cells and mammalian cells in relation to spontaneous mutation rates and in relation to the dose required to cause cell death did not necessarily indicate a similarity in relation to radiation induced mutation rates. In reaching certain of his estimates he appeared not to have realised that monitoring criteria are based on absorbed dose to tissue and not on external radiation. Nor did he appear to have appreciated the importance of the genetically significant fraction of gonad dose in estimating the genetic effects from population exposure. I found nothing in Dr Ichikawa's evidence which could lead me to the conclusion that genetic effects of radiation are significantly greater than presently estimated.

Professor William Potts

10.54 Professor Potts, Professor of Biological Sciences at the University of Lancaster, giving evidence on behalf of the Lancashire and Western Sea Fisheries Joint Committee (LWSFJC), was principally concerned with caesium discharges. He considered that discharges of caesium from Windscale should be limited to 100,000 curies per annum. He was critical, in my view rightly, of (1) the way in which caesium discharges had been allowed to build up to the peak level reached in 1975 without action being taken; (2) the long time taken by FRL to issue their annual reports; (3) the absence of whole body monitoring of individuals in order to test the validity of predictions; (4) the absence of a specific limit of discharge for caesium and (5) the manner in which such limits as there were, were fixed.

10.55 As to the points made, none of them indicate, nor were they intended to, that Professor Potts considered there was any reason why THORP should not be built. He was giving evidence for a sea fisheries committee and was therefore directing his attention mainly to that radionuclide which mainly affects man, whether somatically or genetically, through fish. But he had studied the radiobiological evidence before his own appearance – i.e. during the first 64 days – and he had many years experience in radiobiological matters. Having heard him give evidence, I have no doubt that had he seen any other matter in respect of which he saw any danger to the public he would have raised it. I regard the fact that he raised no other matter as being of considerable significance.

10.56 Professor Potts' criticisms are in my view justified. To a large extent it is accepted that action should and will be taken but since the criticisms affect the system of control rather than the question presently under consideration I will revert to them when I consider that matter.

10.57 His general point that caesium discharges should be limited is met since discharges are intended to be lower than the limit which he suggested.

Dr J K Spearing

10.58 Dr Spearing, a fellow of the Linnean Society and of the Institute of Biology, gave evidence and made submissions on his own behalf. He accepted that the risk to individuals was small but he pointed out that, although this was so, it was certain that radiation would cause some deaths from cancer and also other health damage, not only in the present population but also in future generations. He was principally concerned, however, with genetic effects. His interpretation of, and conclusions from, papers published in recent years was that the harm to both present and future generations was many times grater than presently estimated. He considered that ICRP had ignored the effect of this recent work. He challenged particularly the conclusion, adopted by ICRP, that the application of a linear dose response relationship was likely to over-estimate the effects of radiation at the low doses and dose rates resulting from radioactive discharges in the nuclear industry. In Dr Spearing's view the adoption of a linear dose response relationship would under-estimate the effects of radiation at low doses.

10.59 Dr Spearing's evidence demonstrates another area of dispute among experts. It affords no basis upon which I could conclude that, in issuing their current recommendations, ICRP had ignored the work to which he referred, nor do I see any ground upon which I could hold that, if they did not ignore it, they must have failed to give it adequate consideration. Professor Ellis, to whose evidence I shall shortly refer, and who was a well qualified and impressive witness, regarded present limits as generally cautious and reasonable although he also considered that in certain areas further research was necessary, which research could lead to present limits being either relaxed or made more stringent. I prefer his evidence to that of Dr Spearing.

Professor R E Ellis

10.60 Professor Ellis, Professor of Medical Physics at the University of Leeds and a former member of the Scientific Secretariat of UNSCEAR gave evidence on behalf of TCPA. He had for many years studied radiation hazards and protection and had published some 19 papers concerning these matters. He was concerned principally with the following:
  a. levels of exposure to workers at Windscale, which he considered to be higher than is normally considered acceptable;
  b. levels of exposure to the population living in the vicinity of the works;
  c. levels of exposure to the population consuming locally caught fish or locally produced food;
  d. levels of exposure to the general public.

10.61 With regard to (a) above I find Professor Ellis' criticisms fully justified, not in the sense that the workforce in general is or has been in any immediate danger, but because it was clear that the proportion of workers who had received radiation doses at or near to – and in some cases above – permitted limits was higher than it should be. In the case of THORP, however, the intention is to improve this situation and this can more easily be achieved in a new plant. But it is not sufficient merely to build into the plant improved measures of protection. There is a need for increased attention to procedures for protection and to ensuring their observance. This should be and doubtless will be given attention both by BNFL and NII.

10.62 As to Professor Ellis' remaining points, he was concerned, and rightly, about the levels of caesium discharges and also about the consequences in the event of a 20 fold increase in nuclear power production and thus of reprocessing by the year 2000. The caesium discharges will be very much reduced when the new pond-water treatment plant is in operation and THORP does not, even if run at full throughput of 1,200 tonnes per annum, represent, in terms of reprocessing, anything like the equivalent of a 20 fold increase in nuclear power. As I have already mentioned (para 2.30) the reprocessing needs of the AGRs presently in operation or under construction amount to about 200 tonnes per annum. In the terms of reprocessing THORP thus represents, at most, a 6 fold increase. Since it is intended to operate it for the first 10 years at about 50 per cent capacity the increase will in practice be three fold for the first 10 years i.e. from 1987 (at the earliest) to 1997. At that time the earlier magnox stations will be nearing the end of their useful lives and although use of THORP may then increase there will be a reduction in exposure from magnox reprocessing.

10.63 A convenient illustration of the position from 1987, on the assumption of a 1,200 tonne per annum throughput for THORP, is that the genetically significant dose to the population living within 50 km, from a combination of THORP and magnox is estimated to be about 230 man-rem per annum from caesium 137 and 134, carbon 14, tritium and krypton 85. This population numbers about 300,000. The average genetic dose to this population would therefore be about 23 mrem over 30 years, or one-quarter of the limit of 100 mrem per 30 years which is the limit mentioned in Cmnd 884 for the average for the population of the whole country from waste disposal (see Annex 5). If, therefore, the intended limits are achieved, even the local population genetic exposure will be well within this limit. The contribution to the average genetic dose to the whole population from this source would be very much less: about 0.1 per cent of the limit.

10.64 Professor Ellis' evidence appeared to me to carry great weight for it excluded the tendency, noticeable in the evidence or submissions of some others, to add all suggeste d changes together as if all were established. His evidence does not suggest that THORP cannot be built to tolerable levels. On the contrary it supports the view that it can be so built.

Professor Tolstoy

10.65 Professor Tolstoy was chiefly concerned with the ultimate disposal of waste but I should mention in this context one other aspect of his evidence. He criticised the ICRP standards (for plutonium) as being 'radiation standards whereas we are talking of long-term carcinogenicity (which) leads to far smaller permissible intakes'. He appeared not to have appreciated the significance of the differences between amounts inhaled, amounts ingested, and amounts actually retained in the body; or to have realised that standards were in fact based essentially upon the 'long-term carcinogenicity' attributable to the radiation doses to particular body tissues from this retained fraction.

Dr V T Bowen

10.66 Dr Bowen was principally concerned with what he considered to be the inadequacies of the approach to monitoring of, and monitoring by, FRL. I deal with this subject below. He did however advance two *possibilities* which I ought to mention at this stage:
a. that internal radiation from naturally occurring radionuclides might be virtually harmless i.e. have a threshold, whilst internal radiation from man-made radionuclides had no such threshold; and consequently that man-made radionuclides might have a disproportionately large effect when taken into the body;
b. that the behaviour of americium 241 formed in situ by the decay of plutonium 241 might be different from that of americium 241 released directly in that form.

Both the possibilities were described by Dr Bowen as 'interesting and relevant speculations' although he considered the latter to be 'perceptibly less speculative' than the former. Neither affords any ground for saying that THORP cannot be built to tolerable levels.

Tests made during the inquiry

10.67 I referred in paragraph 1.5 to the fact that certain tests had been carried out during the course of the Inquiry. I include details of such tests at this point in view of the fact that certain objectors sought to rely on the present situation as the basis of their arguments that the building of THORP would create undue risk to the public from routine emissions. A number of suggestions were made as to the existence, or possible existence of alarming situations already present and, whenever it appeared possible that testing of some sort would be likely to provide an indication one way or the other as to the validity of the suggestions made, I asked that such testing should be carried out.

10.68 Before setting out the details it is necessary to stress that such tests were not intended to, and could not, show what was the annual exposure of the public to the particular radionuclide or radionuclides in question. They were intended to give an indication of the likelihood or otherwise of exposures being much greater than hitherto supposed or of a build-up of a particular radionuclide having occurred during the long period during which magnox reprocessing has resulted in routine emissions.

Manchester water supplies

10.69 On 26 July 1977, Mr J Urquhart for the Windscale Inquiry Equal Rights Committee (WIERC) raised the suggestion that two and a half million people in Manchester, which draws its water from the Lake District, might be receiving a significant amount of radioactivity as a result of a build-up in certain lakes of tritium discharged from Windscale. BNFL had in fact been sampling the water of four of the lakes, Wastwater, Ennerdale, Derwentwater and Loweswater for tritium for some years and had found the levels to be below the level of detection. But they had not sampled Thirlmere, Haweswater, Ullswater or Windermere, which are the sources of supply for the Manchester area. At my request all four lakes were sampled by BNFL. By the following day BNFL were able to report the result of a test of Thirlmere. No tritium was detectable. In the time available it would have been possible to detect down to 1/10,000th of the amount presently regarded as permissible in drinking water for continuous use. It followed therefore that such tritium as was present in Thirlmere was less than 1/10,000th of that level. Put in simple terms it would mean that a person could drink something in the order of 10,000 litres of Thirlmere every day without reaching the internationally accepted dose limit. The same position prevailed with regard to the lakes which BNFL had been testing previously. The following day, results from the other lakes from which Manchester draws its supplies and from Coniston Water, Bassenthwaite Lake and Thirlmere (retest) were reported to me. By using longer times for analysis it had been possible to ascertain that in all cases the tritium content was even lower than that first reported for Thirlmere although again no tritium was actually detectable. Samples were also taken from certain lakes and rivers including Haweswater and Thirlmere by the North West Water Authority (NWWA) and analysed for them by the Government Chemist. In all cases the results showed that the tritium present was even lower than the maximum possible levels ascertained by BNFL in the earlier tests.

10.70 In the increased time available the Government Chemist had been able to ascertain levels to an accuracy of ± 20 per cent. The highest level found was about 1/15,000 of the maximum permissible level for drinking water and even this was at the lower end of the range of results obtained from analysis of rainwater in other parts of the country.

10.71 It was not suggested that the present maximum permissible concentration in water (MPC_W) was excessive. I had and have no hesitation in concluding that the Manchester population have no cause for alarm whatever in this matter. They may also be comforted to know that the NWWA regularly monitor for radioactivity in water supplies, and have been so doing since, at least, 1959.

Isle of Man potatoes

10.72 In his evidence for IOM, Dr Bowen raised the possibility of plutonium and americium discharged to the Irish Sea from Windscale returning in significant quantities to man by the following route:—
  i. contamination of seaweed;
  ii. use of seaweed as fertiliser on the potato fields of the Isle of Man;
  iii. uptake of activity by the potatoes;
  iv. consumption of the potatoes by man.

Dr Bowen did not see any cause for alarm, in the sense that he did not think that anyone eating Isle of Man potatoes at present would be exposed to any significant danger, but he was concerned that there had been no monitoring done. He considered it possible that there was a build-up of plutonium or other long-lived transuranics in the soil. He had done no sampling or testing himself but said that he hoped to induce others to do so. In this he succeeded for I requested the IOM Government to have samples of seaweed, soil and potatoes taken and analysed. Such samples were taken and they were tested by NRPB. Their results were made available in the closing stages of the Inquiry. As they were 'one-off' samples they clearly could not indicate whether or not there was any build-up occurring but they did show that there was no cause for any alarm. Counsel for IOM in his closing speech observed, in relation to the results for potatoes, that, in order to reach the internationally recommended dose limit for eating potatoes grown in soil dressed with seaweed for two years a man would have to eat 30 tons of such potatoes per day and, not surprisingly, he expressly stated 'There is no danger from these potatoes'. My own conclusion is the same. It is to be noted, however, that in this case, unlike the case of water from the lakes, there had been no previous sampling. It was submitted that there should have been. I reject this submission. Dr Bowen accepted that no monitoring could test everything and that it must be a matter of judgment whether the possibility that a particular pathway to man might be sufficiently significant to make testing desirable. The results obtained do not suggest that FRL were at fault in not having tested thus far.

Scallops

10.73 The question of sampling the scallops from the Isle of Man for plutonium and americium also arose from the evidence of Dr Bowen. He suggested that only a small revision in the ICRP limits would put the whole scallop fishing industry in jeopardy. In contrast to the case with seaweed, soil and potatoes, the IOM had already asked Dr Bowen's laboratory to analyse samples of scallops but some of the results were not available at the time when he gave evidence.

10.74 When Dr Bowen's final results arrived on the 93rd day of the proceedings they showed that, on the basis of the nuclide concentrations measured, about 50 lbs of scallops could be eaten daily without the ICRP limits being exceeded. Dr Bowen's fear that even a small revision of the ICRP limit would put the scallop industry in danger is therefore without foundation.

10.75 Dr Bowen also raised matters concerning seawater, sand and fishmeal but I find it unnecessary to comment on them.

Air at Ravenglass

10.76 Early in the course of the Inquiry concern was expressed that the public might be facing some significant risk through inhaling alpha emitting particles deposited from the sea on to the mudflats of the Ravenglass estuary and thereafter resuspended in the air. Mr Hermiston gave evidence on 14 July that BNFL had conducted a survey during 1976 using portable air samplers at locations in the estuary down wind of the mud flats. The values measured by BNFL averaged about 3 per cent of the MPC_a for the public for insoluble plutonium in air recommended by the MRC (which is in fact five times lower than that recommended by the ICRP in 1959). The MCP_a is however based on continuous exposure, and Mr Hermiston considered that, since no member of the public was likely to be exposed (if only because of weather and tidal conditions) for more than a few hundred hours per year, the maximum intake by any member of the public was likely to be no more than 0.1 per cent of the dose limit. He therefore concluded that whilst this pathway should be kept under review with a continuing programme of monitoring, there was no immediate cause for concern. This was consistent with the Royal Commission's conclusions on the subject (Sixth Report paras 352–354).

10.77 Mr Hermiston was cross-examined closely on this matter by objectors. It was for example, suggested by Mr D Laxen for NNC that a six month old child living in Ravenglass might be receiving as much as 83 per cent of the ICRP limits. The subject was further pursued with Dr N T Mitchell who gave evidence on behalf of FRL on the 17 and 18 August. He explained that FRL had measured concentrations by means of a different sampling device, the tacky shade, but that no significant differences had appeared between sampling sites near Ravenglass and those further away.

10.78 Since there was considerable dispute about the levels of concentration in the air inhaled by the inhabitants of Ravenglass and suggestions that they were or might be very much higher than believed, I invited all parties concerned to see whether they could agree with the relevant Government bodies on some form of practical monitoring which, before the close of the Inquiry, might provide a reliable guide to the exposure of the inhabitants of Ravenglass through inhalation. All parties were properly concerned that the basis of any monitoring carried out should be scientifically sound, and initially they took the view that nothing which could be done within the time available would be of any real value. I was not however convinced that this was so and requested the parties to give further consideration to the matter. The result was that all parties agreed on a programme of sampling which it was considered would be of value. On the 23 August Mr Morley of the NRPB informed me of the details as follows:

'Sir, Board staff have consulted with the scientists of the interested parties, and further consideration has been given to your suggestions relating to the additional sampling of dust in Ravenglass. We have taken into account your statement that you wished to have information which would be valuable but not an accurate estimate of the annual dose. All parties have agreed that the most useful information which could be produced on a timescale of three or four weeks would be obtained by sampling the airborne dust outdoors in Ravenglass using high volume air samplers. The samples taken would be analysed for americium 241 and plutonium 239 plus plutonium 240. The results obtained can be compared with appropriate maximum permissible concentration values to provide an indication of the significance of the concentration of these nuclides in the atmosphere, but only during the period in which the samples are collected. No useful additional information could be obtained on a short-term basis from samples of household dust or road dust.

'We therefore recommend to you, sir, that the short-term measurement programme should be confined to the sampling of air-borne dust outdoors. The programme suggested could begin this week. It is proposed that two samplers should be used, one at each end of the main street in Ravenglass. It is also proposed that a similar sampling operation should be undertaken to serve as a control for reference purposes, and for convenience this will be located at a site near Harwell chosen in such a way that any radioactive emissions from the Harwell site will have no effect on the analytical results obtained. The sampler filters will be changed either daily or every few days. The time required for analysis will be 12 to 15 days after each filter has been removed. At Ravenglass a note will be kept of wind, weather and tidal conditions throughout the sampling period. It has been agreed it will be left to the Board to arrange for the analyses to be undertaken in the Board's laboratories or the laboratories of the Environmental and Medical Sciences Division at AERE, according to the availability of resources. The laboratory of the Government Chemist can assist if necessary.'

I requested NRPB to proceed with the programme as agreed, to keep the sampling going for a full month if possible, and to report interim results.

10.79 Sampling commenced on 26 August and continued until 22 September; seven written reports were sent by the NRPB to the Inquiry at regular intervals after analysis of the results. In addition an interim oral report with written summary of results to the 9th September was made by Mr O'Riordan on 30 September and a final oral report with written summary of results over the whole period was made by Mr Shaw on 17 October. The results of the exercise are summarised in the table below. The mean and highest measured concentrations from each of the two sites are expressed as percentages of ICRP maximum permissible concentrations in air for the general public and, in the case of insoluble plutonium, also as percentages of MRC recommended level of 1/5th of ICRP.

	Northern Sampling Site		Southern Sampling Site	
Plutonium 239 and 240	Mean result (28 results)	Max result	Mean result (28 results)	Max result
%ICRP MPC$_a$ Soluble	0.37	1.22	0.10	0.25
%ICRP MPC$_a$ Insoluble	0.022	0.073	0.006	0.015
%MRC MPC$_a$ Insoluble	0.11	0.37	0.03	0.08
Americium 241				
%ICRP MPC$_a$ Soluble	0.09	0.32	0.019	0.06
%ICRP MPC$_a$ Insoluble	0.005	0.02	0.001	0.003

It will be seen that there is considerable variation between the levels recorded at the two sites and between the mean results and the maxima at each site. It was not possible to investigate the precise causes of these wide variations but one cause of the higher levels at the Northern site was probably the fact that the samplers were placed close to a point where buses turn and raise deposited dust. Variations between results at the same site are not surprising for there was, over the period of the sampling, a wide range of weather conditions.

10.80 It will also be seen that the mean results would still be within ICRP limits even if such limits were reduced, in the case of soluble plutonium by a factor of about 300; in the case of insoluble plutonium by a factor of about 4,000, in the case of soluble americium by a factor of 1,000 and in the case of insoluble americium by a factor of 20,000.

Although the results include figures for soluble plutonium and americium this is for completeness only, for it is difficult to see how any significant part of either radionuclide could, if soluble, reach the air samplers in any significant quantity.

10.81 Provided that the results can be relied upon as giving a reasonable indication of the concentrations likely to prevail during the course of a year and provided that the recommended MPC$_a$'s are not grossly overstated it is plain that the inhabitants of Ravenglass need have no fear that there is any present risk to health.

10.82 Since the results were obtained by daily sampling over a period of a month in the summer I consider it to be unlikely that sampling over a year would show, on average, markedly higher concentrations. This was the view expressed by Mr Shaw of NRPB and I accept it. Whilst some revision of secondary limits for particular radionuclides may undoubtedly be made, I neither heard nor read any evidence which gave me any cause to believe that present limits are seriously in error. I therefore reiterate the opinion which I thought it right to express in the course of the Inquiry that the inhabitants of Ravenglass need have no cause for alarm. This view was expressly accepted by Professor Ellis and, in more general terms, by Professor Potts. Neither of these witnesses was concerned to promote the interest of the nuclear industry, nor were they supporters of the application.

10.83 NNC submitted through Professor Radford, Dr Wynne and Mr Laxen that there was cause for alarm. I have already considered Professor Radford's evidence and it is unnecessary to say any more about it. It is also unnecessary to say very much about the propositions advanced by Dr Wynne and Mr Laxen. Dr Wynne considered that no assurance could be given to the inhabitants of Ravenglass until an extensive five-year monitoring programme had been completed. Mr Laxen also took this view and appeared so determined not to acknowledge that the results of the monitoring warranted an assurance that he became somewhat fanciful. I am unable to attach any weight to their submissions.

10.84 There is no doubt that the situation at Ravenglass should be kept under review and BNFL intend to do this. It would however be a disservice to the public to divert resources to a massive monitoring programme which, if it were to satisfy Mr Laxen, would require samplers to be placed almost everywhere.

Whole body monitoring of fish eaters for caesium

10.85 The most significant discharges from Windscale for some years have been those of caesium 134 and 137. FRL's report for 1976 showed calculated exposure of the maximum fisheater who was estimated to consume 265g per day of fish caught in the vicinity of the Windscale pipeline as 44 per cent of the ICRP limit. No validation tests had been carried out by way of whole body counts for caesium on individual fish eaters as part of FRL's monitoring programme. It was submitted by objectors and accepted by BNFL and FRL that validation tests would be desirable although it was considered that, if undertaken, they would be likely to show that exposures had been overestimated rather than underestimated. It was also accepted from the outset that caesium discharges were higher than desirable.

10.86 Friends of the Earth (West Cumbria) (FOE WC), whose case was presented by Mr C Haworth, called a number of local witnesses, one of whom specifically raised the question of harm which might result from eating locally caught fish. This witness was not in fact an eater of fish, but another witness called by FOE WC, Councillor W Dixon, stated that he was a regular fish eater and it appeared to me that, if he was prepared to allow himself to be subjected to whole-body monitoring for caesium this might enable tests to be initiated without unnecessary alarm. I therefore invited him to do so and he readily agreed. Having done so he volunteered the view that it would be desirable if more people were monitored in this way. As a result, with the co-operation of Mr Haworth, Councillor Dixon and Mrs M Higham, an individual objector, 17 volunteers were obtained and a programme of monitoring was agreed between BNFL, Councillor Dixon, Mr Haworth and Professor Fremlin. Councillor Dixon was himself one of the volunteers. The others must remain, at their own wish, anonymous. I have already mentioned my thanks to all concerned (para 1.5 above). I repeat those thanks here, for not only did their co-operation enable valuable information to be contributed to the Inquiry, it also enabled a start to be made on validation testing which everyone agreed should continue. I return to this later.

10.87 The agreed programme was duly carried out and the results were set out in a written report (BNFL 326) handed in on the 27 October. I do not set them out in detail. The volunteer with the highest body content had consumed 33 oz of fish per week and it was estimated that his exposure from continuous consumption at this rate would be about 8 per cent of the ICRP limit. Mr Haworth stated with regard to the results:—
> 'As to the figures themselves, sir, I was particularly anxious that this should appear as scientific a document as possible, which I think it does. There are no value judgments in it whatsoever and therefore it is completely acceptable to me as a document so that the interpretation of these by an individual is entirely a matter for their own judgment. My own judgment is that while the levels here are certainly, as is stated, not going to give grounds for immediate concern, I don't think to anybody, this is a situation that needs to be watched.'

This statement displayed a balanced moderation which was typical of the whole presentation by Mr Haworth of the case for FOE (WC) and indeed of the local witnesses called by him. I can find no better place than here to express my appreciation to him and to them for a very valuable contribution to the Inquiry.

10.88 My own judgment accords with that of Mr Haworth. The results show that to reach the ICRP limit the most exposed of the 17 volunteers would have to consume almost 2 stone of fish per week.

10.89 Since the point was specifically raised by one volunteer I should make it plain that no volunteer, by submitting to the monitoring and to the publication of the results, should be thought thereby to have agreed that any amount of radiation from fish or any other source is acceptable. Volunteers did no more than agree to the facts concerning their body contents of caesium being ascertained and supplied to the Inquiry.

10.90 The results also provide a useful comparison with FRL's calculations for exposure of the critical group in 1976. These calculations showed, as already mentioned, an exposure of 44 per cent ICRP dose limit on a consumption of 265g per day of fish caught exclusively within 5 km of the terminus of the Windscale pipeline. Had the consumption by the most exposed volunteer been about the same as that assumed by FRL, that volunteer would have reached only about 16 per cent of ICRP dose limit. Had that volunteer eaten as much as 830g per day (a figure which it was suggested FRL should have taken for the maximum consumer) he would still have reached only 50 per cent of the ICRP dose limit. The difference may be accounted for by the fact that, whilst FRL employed the maximising but unrealistic assumption that all fish eaten were caught within 5 km of the pipeline, volunteers consumed fish from a wider and more realistic area. It is also useful to consider the comparative situation if, instead of taking the volunteer with the maximum consumption only, the average exposure per oz of fish over the 17 volunteers is used. On this basis a consumer of 830g of fish per day would reach an exposure of about 61 per cent ICRP dose limit.

10.91 In the light of the results found I reject the suggestion that FRL have been under-estimating exposures. It appears to me that they have, very properly, estimated on a basis which provides a considerable margin of safety. There is, I accept, an under-estimation if they should have taken their maximum eater as consuming 830g per day of fish, all of which was caught in the immediate vicinity of the pipeline. I do not accept that they should have done so.

10.92 As regards the future, it is accepted that whole body monitoring of volunteers from the local population should be continued. Many of the fears expressed to me by local people giving evidence were the result of a lack of information, and they should clearly be in a position to reassure themselves that all is well.

10.93 I do not consider that any planning condition could be attached on this matter in the event of permission being given for THORP and in any event it was accepted that more permanent arrangements should be instituted whether or not planning permission is given. The details are in my view best left to be settled by agreement between BNFL, the controlling authorities, and the local people and their representatives on the local councils. I would however suggest that the following general principles should be adhered to, whatever the detailed arrangements:—
  a. Monitoring of members of the public should be on an entirely voluntary basis; although I appreciate that some encouragement may need to be given

if sufficient numbers are to volunteer.
   b. The monitoring facility should be open free of charge to any member of the public who wishes to be monitored.
   c. In all cases the results of the monitoring and their meaning should be explained to those concerned.
   d. Should any member of the public so wish, the explanation of the results should be given by an independent and authoritative person not connected with BNFL. This person should also be available for consultation with any member of the public, whether monitored or not, who wishes to seek information or reassurance about the effects of radioactive discharges from Windscale. In suggesting that this person should be independent I wish in no way to cast doubt on the integrity of BNFL's medical staff. Plainly, however, if the public are concerned because of fears about BNFL's activities they will place more reliance upon professional advice from somebody completely independent of it.

10.94 So far as the provision of whole body monitoring facilities are concerned Counsel for BNFL announced in his closing submission that the company had it in mind, whatever the decision on THORP, to provide an additional whole body monitor for use by members of the general public. Although it would be within the site perimeter, no problem of access would arise provided an appointment were made, as is the case with any other form of medical examination. No doubt this proposal will be generally welcomed, but I would suggest that further thought be given to the location of the facilities. Even at the best regulated works, things can go wrong. It is likely that the public will be most anxious to use the facilities in the event of some mishap having occurred. Depending on the scale of the event, it might indeed be essential that the facilities were then used. If they were located within the works boundary, however, they might have to be quarantined as a result of the event, and, even if this were not the case, the public might be understandably reluctant to use them. Welcome, therefore, though BNFL's proposal is, I would recommend that serious consideration be given to finding a site away from Windscale for locating these facilities.

10.95 It would not be appropriate for me in this report to make any more detailed recommendations on the arrangements to be adopted. It was however brought to my attention that the authorising departments would have the power to attach a requirement that facilities be provided as a condition of the authorisations to discharge. They will no doubt wish to consider whether, in the light of the action which has already been voluntarily undertaken, such a condition would be desirable.

Blistered fish

10.96 Neither this matter nor that which follows involve tests but both require mention somewhere and I include them under the heading of tests for both are cases where alarm about a particular matter was expressed. On 12 October Mr Stredder, an itinerant street entertainer, reported to me a conversation he had had with a local fisherman to the effect that fishing had been curtailed in certain parts of the Irish Sea because of blisters found on fish and thought to be caused by radioactive pollution. At my invitation the County Council made enquiries of the relevant sea fisheries committees (the Cumbrian and the Lancashire and South Western).

10.97 The replies received showed that there were no such restrictions, that blistering of fish due to a viral disease known as Lymphocystis was well known and was more prevalent in the Irish Sea than in other home waters, that the disease in the Irish Sea had been reported on as long ago as 1904 and 1905, that it was not considered to be attributable to Windscale (not surprisingly in view of the last two dates) and that it was presently on the decline.

10.98 This matter is a particularly good instance of the readiness with which anything will be attributed to radiation. The fact that the phenomenon was known to be particularly prevalent in the Irish Sea some 45 years before any nuclear operations began at Windscale and 40 years before Hiroshima should convince even the most anxious that in this case at least Windscale is not guilty.

Radioactive furniture

10.99 On 2 November, the 98th day of proceedings, I heard an application from Mr Miller who was concerned that his daughter was being put at risk by using articles of furniture previously in the ownership of a man who had been contaminated by plutonium. I explained to him that the NRPB were in certain cases prepared to, and had in this case agreed to, carry out the necessary tests, and I urged him to make the necessary arrangements. At the time of writing, I am informed that no approach has been made to the NRPB.

10.100 Having dealt with the individual matters I conclude with a general observation. I accept that, since Ravenglass sampling only occupied one month, as did the fish eating programme, and since other results were from one-off samples, no accurate estimates of exposure can be made from them. When, however, one finds that preliminary explorations of a number of suggested sources of danger produce no single instance showing either immediate cause for alarm or any evidence of a build-up after 20 years of discharges, it appears to me to afford good grounds for reassurance. If, when many people are looking for possible areas of danger, I find (a) that in every case save one there appears to be a margin of safety so large that present limits could be radically changed and yet still leave a large margin of safety and (b) that in the one case where there is only a safety factor of two, this is well known to the authorities responsible for protecting us and BNFL are engaged on installing a plant which will radically improve the situation, I am encouraged to think that there has not been, is not and is not likely to be any real cause for alarm.

Is the system defective?

The international aspect

10.101 The Royal Commission concluded (para 527 – conclusion 3) that 'there is no better way of deriving basic standards than on ICRP recommendations, given that the scientific standing and independence of its members is maintained'. This conclusion was challenged by a number of objectors. It was generally accepted that ICRP had done much valuable work but it was submitted
  i. that although fully independent it was, being in effect a largely self elected body, unlikely to elect a person, however well qualified, whose conclusions ran counter to those of its members. It would therefore be opposed to change. This was shown by the fact that basic standards for whole body exposure had not changed since 1959 and were in fact confirmed in ICRP26.
  ii. that it did not include in its membership sufficient geneticists.
  iii. that it had given inadequate attention to recent work, or had reached incorrect conclusions on such work.
  iv. that in any event it should not both assess risks and recommend limits. It should be limited to assessing the risks. Limits should then be recommended by another body because the limit to be adopted depended on factors which were not matters of scientific expertise.
  v. that their publications were of little weight because they had not been properly refereed.

10.102 I accept that a largely self-elected body may tend to perpetuate its own thinking but it need not do so and the fact that standards have remained unchanged is no evidence that ICRP has done so. It is neutral. Standards would remain unchanged if they were correct when last recommended or if they were then too strict but the Commission considered that, since it had proved possible to comply with them, it was unnecessary and undesirable to relax them. The fact that, despite papers suggesting that the last recommended limits were much too high, the Commission adheres to those limits again means nothing. The Commission may have considered such suggestions and concluded that they were invalid. It may have been right in so doing and it may be that, had it accepted and acted upon the suggestions it would have provoked a storm of criticism from the scientific community far exceeding the criticism which had stimulated their action. ICRP may indeed be described as a sitting duck. If it does not change a standard it can be accused of not taking into account modern work. If it varies a standard upwards it can be accused of seeking to further the interests of nuclear power at the expense of public safety. I have little doubt that if the Commission had been excluding people who, on scientific reputation, should have been included, there would have been complaint in some quarters. If, in the future, the Commission's elections revealed that it was excluding people who ought to have been included I have equally little doubt that the fact would promptly be noticed. With the international and national checking that is in constant operation there are, in my view, adequate checks for the protection of the public.

10.103 As to the alleged insufficiency of genetic expertise, ICRP's committee No 1 on radiation effects seems to me to be the body on which there should be adequate representation by geneticists and at the time of publication of ICRP26 three of its 12 members were geneticists. This can scarcely be termed inadequate representation.

10.104 Before passing to the fourth of the points mentioned in paragraph 10.101 above I should mention the question of the secondary limits. As previously stated there were suggestions that the current limits relating to plutonium should be very greatly reduced. This matter was considered by the Royal Commission who concluded that with the reduction already recommended by the MRC in respect of the MPC_a for insoluble particles the secondary limits were not seriously in error (Sixth Report paras 77 and 526 Conclusion 2). This reduced limit is already operated in the UK. ICRP committee 2 are currently completing a review of secondary limits with a view to updating their current recommendations. They may or may not alter some or all of existing limits. It is however a matter for them.

10.105 The suggestion that ICRP should merely assess risks appears to me wholly misplaced and liable to lead to confusion. I accept of course that scientists have no special expertise as to the question of the risk level which the public at large would consider tolerable but if, as is the case, some body has to consider the question, it appears to me both that ICRP is as well able as any other body to do so, and that to divide the two functions is merely to proliferate committees to no useful purpose. This is particularly so when it is remembered that what is ultimately the limit operated in any country is a matter for each individual country. It may well be that one country would find the tolerable level of risk only one-tenth of ICRP limit. If it does so consider, it is free to impose that limit.

10.106 The fifth suggestion I confess to finding a little absurd. ICRP's publications appear to me to be subjected to the severest possible scrutiny and the suggestion that they would be the more authoritative if they had appeared, after the ordinary refereeing process, in some scientific journal is to my mind quite untenable. In fact a number of the reports of its committees have appeared in Health Physics and other journals. It does not appear to me to be reasonable to suggest that such reports should be regarded as more authoritative than those which ICRP publish directly.

10.107 Although I agree with the Royal Commission's conclusion quoted in para. 10.101 above and have no reason to suppose that any variation of the system in its international aspects is required there is one matter upon which it may be useful to comment. At page 47 of ICRP 26 the Commission described its method of work in the following terms.

> 'Much of the work of the International Commission on Radiological Protection is performed by ad hoc task groups, by means of which the Commission has been able to call on the services of a large number of individuals who are not members of a committee. In this way the Commission is able to bring together the appropriate experts rapidly and effectively so that work can be completed and reports published within the short time that is essential if the Commission's recommendations are to ensure the safe and rapid development of new techniques in the light of the most recent information. Reports that have been published are shown at the end of the appendix.'

Although the Commission recognise that speed is essential and consider that the essential speed has been observed, its recommendations as to basic standards appear infrequently. I do not suggest that there has been any need for more frequent recommendations but it does appear to me that some form of interim reporting would, if practicable, be desirable. In his evidence Dr Bowen said:

> 'Unless our attention is directed not merely to maintaining safe operating practices, but to convincing most of the public that we know these are safe and helping them understand how we know these are safe, then there is a very real possibility of democratic refusal of further development of our nuclear potential.'

10.108 The evidence at the Inquiry convinced me that this observation carries weight and I have no doubt that understanding would be increased if ICRP were to issue some form of interim reports. I take by way of example the Mancuso, Stewart and Kneale paper. This will no doubt be considered – indeed it may well already have been considered – by the appropriate committee of ICRP as a matter of some urgency. It may be that, having considered it, they will for good reason reject its conclusions. If this happens there will be no need for further recommendations but if silence reigns it will enable doubt to be raised in the public mind whether ICRP have considered it. I do not suggest that there is any need for ICRP to publish annual reports with full reviews of all the work it has considered but if it were practical to publish a list of papers to which consideration had been given in the previous year it would at least prevent suggestions being made that current work had been overlooked.

The national aspect

10.109 As already indicated, (paragraphs 10.6 to 10.8 above), the first step in the national system as it will now operate will be for the Secretaries of State for the Environment and for Scotland and Wales to consider ICRP/EURATOM recommendations and, with the advice of NRPB and MRC, to decide whether such recommendations should apply in the United Kingdom and, if not, to fix more stringent standards. I neither heard nor read anything to convince me that such a procedure would be unsatisfactory or inadequate and it has the advantage that there will be a straightforward Ministerial responsibility to Parliament for action taken or not taken.

10.110 It is however of prime importance both that, as the Royal Commission stressed, NRPB should be seen to be independent and that the basis of the decision taken should be fully revealed. I do not doubt that NRPB are in fact independent nor do I understand the Royal Commission to have expressed any doubts on this matter, but there is clearly an atmosphere of suspicion in certain quarters. Such suspicion may be inadvertently fostered if, for example, as in fact happened, NRPB, having been in correspondence with a member of the public, forwards copies of the correspondence to BNFL with the comment that they are unlikely to have much trouble with the person concerned.

10.111 Explanation of the basis of the decision whatever it may be is essential. There was, at the Inquiry, much talk of public participation. This I think is unreal. Consideration should however be given to formal inclusion in the advisory process, not merely of NRPB and MRC, but of some wholly independent and responsible person or body with environmental interests. That such a person's or such a body's approval of standards would do much to ensure that the public did not suffer needless anxiety appears to me to be beyond doubt. The acceptance, for example, by one of the environmental groups or by an independent expert such as Professor Fremlin of the safety of any set of standards would be far more reassuring than any pronouncement by a Government body however independent it might in fact be. Should it be decided to bring an independent body or person into the process, it would probably be necessary to change such body or person from time to time in order to avoid suggestions that prolonged participation had undermined independence.

10.112 The next matter is that of control of discharges. The Royal Commission, whilst satisfied that the present control arrangements had worked reasonably satisfactorily, recommended that, for the future, a single new inspectorate should be responsible for determining and controlling all discharges to the environment in consultation with MAFF and NRPB. (Sixth Report – para 527 conclusion 6). The new inspectorate suggested by the Commission was that which had been recommended in its Fifth Report, namely HM

Pollution Inspectorate or HMPI. In the White Paper (Annex 2) the Government reported that it had not yet reached a decision on the recommendation in the Fifth Report that there should be a unified pollution inspectorate and was not therefore in a position to express a view on the recommendation that the responsibility for controlling discharges should lie with such a body.

10.113 The evidence given at the Inquiry led me to share the conclusion that present arrangements have worked reasonably satisfactorily but it also left me in no doubt that a single body would be an improvement on the present situation, for there exists a degree of uncertainty as to where respective responsibilities lie. I would however go further. It would not only be an improvement, it is in my view necessary in the public interest for three simple reasons, because:
 a. the public may be, indeed to some extent inevitably will be, exposed to radiation from more than one source. A man may, for example, receive radiation doses, none significant in themselves, from the fish which he eats, the water or milk which he drinks, the air which he breathes, the ground on which he stands whilst working and the materials of which his house is constructed. It appears to me, therefore, that an overall view must be taken when the question of the level of discharge to be authorised is considered. This no doubt *can* be done by consultation if there is divided responsibility, but a divided responsibility is less likely to ensure that it *is* done. At past and present levels this aspect has not been of particular importance but it could easily become so.
 b. a discharge to water may, e.g. by subsequent deposition and resuspension, reach man via the atmosphere, whilst a discharge to the atmosphere may, e.g. by condensation, reach man via water.
 c. where an authorisation is being considered, it is important that possible consequences of alternatives should be taken into account. It might for example be desirable, from the single standpoint of radiological protection, to restrict discharges from a proposed project to a point which could not be economically achieved. The restriction, if imposed, might thus prevent the project going forward. This would, however, be of little benefit to the public if it resulted in an alternative project which would do greater harm or create greater risk of harm. A unified pollution inspectorate would have the necessary information to take such matters into account.

10.114 Arising out of sub-paragraph a. of the preceding paragraph it should be appreciated that the process of control of discharges must involve radiobiologists. When levels of radiation are very low, and virtually all radiation comes from one installation this matter may be of comparatively little importance; but, if levels rise and the sources multiply, the control process becomes more complicated. Suppose, for example, that the question at issue is the level of discharge to be permitted to sea from a new installation somewhere on the east coast. Looking only at discharges from that installation the critical group might appear to be a group of fish eaters on the coast. If, however, radiation from other sources were taken into account the position could change or it might be found that what appeared to be safe for that group was not safe at all. Hence an essential step in fixing limits will be to ascertain levels of radiation already existing.

10.115 Discharge authorisations in the past have, in the case of aqueous discharges, not specified limits for each radionuclide. The authorisation from Windscale in force since 1970 only specified limits from strontium 90 and ruthenium 106. Everything else was covered by a block limitation for alpha activity and beta activity. In the case of atmospheric discharges there is not even a block limitation. This is accepted as being unsatisfactory. Had there been a specific limitation for discharges of caesium 137 at the time when discharges escalated it is, I think, very probable that action would have been taken sooner. In the future there should in my view be specific discharge limits for each significant radionuclide whether the discharge is to sea or to atmosphere.

10.116 The process by which limits are fixed also requires improvement. At present, so far as aqueous discharges are concerned, limits are fixed by a process of negotiation in which it appeared to me that the public interest, although protected, was given insufficient emphasis. That interest requires that all discharges are kept as low as reasonably practicable and the authority should put the onus squarely on the operator to show that a discharge cannot practicably be avoided. There is, at present, a tendency either to ask the operator what he proposes to do and to accept it if the result is within the levels regarded as permissible, or to suggest a limit, in which case the operator may accept it although he could at comparatively little cost have kept it to a lower level.

10.117 The minimisation of discharges is important not only in itself but also in order to allow for additional discharges if and when the industry expands. In this connection it is important to observe that, in this country, we do not fix national dose limits according to type of practice, as is the case for example in the USA. The overall limits recommended by ICRP and adopted by EURATOM are limits for the total of all practices (except medical). Hence the control authorities must, when fixing limits for any particular discharge, allow margins not only for other sources of radiation exposure of the same groups of people but also for radiation which may be added in the future.

Public participation in the control system

10.118 Before passing to the question of monitoring I draw attention to an important aspect of the control system. I have already mentioned that discharges of radioactive waste require a joint authorisation from the

Secretary of State for the Environment and the Minister of Agriculture, Fisheries and Food under the Radioactive Substances Act 1960. Under Section 8(2) of the Act each of the Secretary of State and the Minister is obliged, before granting an authorisation, to consult with such local authorities, river boards, local fisheries committees, statutory water undertakers or other public or local authorities as appear to him proper to be consulted by him. Under Section 8(4) limitations and conditions may be attached to the authorisation. Under Section 8(5) a copy of the certificate of authorisation must be provided to the local authority in whose area the waste is to be disposed of and to any other public or local authority consulted under Section 8(2), unless it is considered that for reasons of national security it is necessary that knowledge of the authorisation should be restricted.

10.119 It is clear from the foregoing that those representing the public interest will be consulted before the grant of an authorisation and will, in the ordinary case, be informed of the details of the grant when the decision has been made. The matter does not however end there. Under Section 11(1), before an application for an authorisation under Section 6(1) is refused or is granted subject to limitation or conditions, and before any authorisation is varied otherwise than by revoking a limitation or condition, the applicant or the person to whom the authorisation was granted must be afforded an opportunity to appear before and be heard by a person appointed for the purpose. The Secretary of State and the Minister may also afford to such local authorities or other persons as they may consider appropriate the same opportunity to appear and be heard.

10.120 The foregoing provisions provide additional opportunities for representations by or on behalf of the public but they are discretionary only. Moreover it appears strange that, whilst the applicant has a right to be heard on a proposed refusal to authorise, on any proposed limitations or conditions and on any proposed variation of limitations or conditions, no-one representing the public has a right to be heard on a proposed grant or on proposed limitations or conditions otherwise than by consultation under Section 8(2), and that no-one representing the public has even the right of consultation on a proposed variation of limitation or conditions. Indeed, if the proposed variation consists of the revocation of *all* conditions and limitations, there is not only no right to be heard, there is not even power to afford an opportunity to be heard. All that will happen is that such public or local authorities as have received copies of the original authorisation will receive notice of the revocation of the conditions under Section 8(8).

10.121 There appears to me to be an imbalance in these provisions. It can be said that public protection is sufficiently afforded by the authorising departments themselves but this, while no doubt a good answer in many fields, is not in my view satisfactory in this field. I would suppose that, only in rare cases would the authorising departments revoke a condition without consultation or fail to exercise the discretionary power, where it existed, to afford an opportunity to local authorities to be heard on a variation of conditions. But this is not an answer likely to appeal either to the local authorities concerned or to the public whom they represent. Moreover the provisions are illogical in themselves. Suppose the departments propose to revoke completely a limitation on discharge to sea of, say, caesium 134 from Windscale. There is no power to hold an Inquiry at all, and no obligation to consult local authorities. If however the proposal were to double or halve the authorised discharge then BNFL would have a right to be heard but Cumbria, Copeland, the NWWA and the Fisheries Committees would have no such right. I am unable to see any good reason why this should be so. If the proposal were to double the authorised discharge Cumbria, or any of the others, might wish to contend that the increase should not be permitted at all, or that a lesser increase should be allowed or that, far from being increased, the existing authorisation should be decreased. If, on the other hand, the proposal were to halve the existing authorisation BNFL, exercising their right to be heard, might put up an apparently powerful case to reduce it by only 10 per cent but Cumbria would have no right to test the case made or to advance a case of their own that, for example, the decrease should be greater than proposed. They might be afforded an opportunity so to do but it appears to me that it is unsatisfactory that their ability to do so should depend upon an invitation.

10.122 These provisions should, in my view, be re-examined. Curiously they did not attract the attention of any objector, despite impassioned pleas for greater public participation which might have been better directed to suggestions for the improvement of these provisions than to the less substantive targets selected. The provisions appear to recognise the need for public participation in the only sense in which it is real, that is through elected representatives, but to have failed to some extent to afford it. I do not suggest that on every grant or variation there should be a public Inquiry. In the majority of cases this would no doubt be unnecessary. I do however recommend that local bodies be put on an equal footing with the applicant and that the possibility of revocation of conditions without either consultation with or representations by local bodies be removed. Had the matter been argued by objectors I might have been in a position to be more specific. As it is I can go no further.

Compensation for harm done by radiation

10.123 It was submitted on behalf of FOE and others that the provisions of the Nuclear Installations (Licensing and Insurance) Act 1959 and the Nuclear Installations (Amendment) Act 1965 were wholly inadequate to enable compensation to be recovered by a person who, notwithstanding the small risk of harm, did in fact suffer harm from radioactive emissions.

10.124 Those acts were clearly intended to enable such a person to recover compensation and, by extending the period of limitation to thirty years, equally clearly recognised that the damage suffered might not occur or become apparent for many years after exposure to the radiation giving rise to the claim. The apparently beneficial effect of the provisions is, however, likely in the ordinary case to be illusory. By the ordinary case I mean radiation induced cancer. Since, at low levels of radiation, the risk that any particular individual will die of a radiation induced cancer are small and since it is impossible to determine whether a cancer is radiation induced or natural, it follows that it will be virtually impossible for a person who has been subjected to radiation and has contracted a cancer to establish that the cancer was due to radiation and thus that he is entitled to compensation. By way of illustration it is convenient to consider the case of a man who was employed at Windscale say from 1960–1980. In 1987 he dies of lung cancer aged 72. He was exposed to radiation at say 1 rem per annum for 20 years but he is a regular but moderate smoker. He is thus far more likely to have contracted his cancer from smoking than from exposure to radiation and his claim will therefore fail even though his cancer was radiation induced. Indeed even if he is a non-smoker he must fail, for at 72 the chances of dying from cancer from some other source will be greater than the chances of dying of cancer from the radiation exposure.

10.125 It was suggested that in order to give proper protection it would be necessary to have some provision that anyone dying of cancer who could show that he had been exposed to certain levels of radiation should be presumed to have contracted his cancer from that radiation. This would, however, without some qualification, result in the operator of the installation paying compensation to numbers of people who certainly had not died of radiation exposure and to others who probably had not so died. A provision, which, for example, limited the presumption to those cases in which the sufferer was not a smoker, might alleviate the position but it would still remain the case that compensation would be paid in cases where the radiation was not the culprit.

I report the matter since it was raised and so that it may be considered. The problem is whether the law should be so radically changed that, in the special case of radiation, whether from THORP or elsewhere, a claimant should be entitled to recover damages on mere proof that he could possibly have suffered harm from the cause giving rise to the claim, instead of being obliged to establish that the harm suffered was probably due to such cause. I feel unable to recommend any such change.

Monitoring

10.126 It was suggested that the body which fixes the limits should not also monitor results. I see no merit in this suggestion. A body which has fixed the limits will be concerned to see that they are achieved. It may be that, although the discharge limits are achieved, the resulting doses are found to be higher than expected, but I see no reason to suppose that, if this occurred, the authorising body would fail to take action or would try to conceal the results. There have however been three clear defects in the monitoring system to the present time. It does not include validation tests of predicted results by, for example, body counts on exposed persons. This matter I have already dealt with (para 10.93). The second defect is the apparent inadequacy of monitoring of atmospheric discharges. BNFL conduct their own monitoring but there appears to have been little monitoring by any official body. There clearly should be. The third defect lies in the long delays which have occurred in the publication by FRL of its annual reports. Spurred on by the Inquiry there was a marked acceleration and FRL accepted that their reports should be published much more rapidly in the future. There should clearly also be an annual survey to cover discharges to the atmosphere and land. I agree entirely with the Royal Commission's recommendation that there should be one comprehensive annual survey and also at intervals a report by NRPB on radiation exposure.

Research

10.127 The Royal Commission summarised, in paragraph 528 its conclusions and recommendations with regard to research. The Government response is set out in paragraph 25 of, and paragraphs 2, 11 and 12 of Annex A to, the White Paper. The evidence given at this Inquiry gives me no cause to add anything to what was said by the Royal Commission.

10.128 The principal matter raised at the Inquiry was the need for independent and parallel research to be carried on. Whilst I agree that this is important it was clear that a vast amount of independent research is carried on and has been carried on for many years. Many of the papers handed in were by independent persons or bodies, many such papers were the joint work of persons belonging to official bodies in collaboration with others who did not so belong, and even a cursory look at the lists of references in the papers handed in shows how much more material there is which is of an independent nature.

10.129 I reject completely any suggestion that research and its results have been kept within the confines of those institutions whose interest it is to promote the growth of the nuclear industry or to control its operations.

Quality and integrity of the advisory and control authorities

10.130 I reject completely suggestions made that the control institutions were serving the interests of the nuclear industry in disregard of the public and the workforce. Having heard witnesses from FRL, NRPB, NII, DOE, UKAEA and the Department of Transport (DTp) I have no doubt as to the integrity of those concerned in all of them and I regard the attacks made on them as being without foundation. Such attacks did nothing to further the cases of those who made them and at times reached a level of absurdity which was positively harmful to such cases. It is noteworthy that no attacks of this nature were made by FOE.

10.131 Much stress was laid on the fact that some of those whose duty is to serve the interests of the public have formerly been employed within the industry. I can see no disadvantage in this. Indeed it appears to me to be a clear advantage. Such persons bring with them a knowledge from within the industry which can only be of advantage when they join a control organisation.

10.132 If the control organisations were to be deprived of persons coming from within the industry because of a possibility – which I accept exists – that one or more may have a pre-existing bias sufficient to impair his ability to look after the public interest, it would seriously weaken their effectiveness. Moreover it must be remembered that a pro-industry bias may in some circumstances lead to a person advocating even greater precautions than those suggested by others with no such bias in order to ensure that there can be no excuse for accusing the industry of harming the public.

10.133 As to the capability of the control or advisory institutions FRL were subjected to the most stringent criticisms of a far reaching nature and NRPB were also severely criticised principally in connection with the paper by Dr Dolphin comparing observed and expected cancer deaths of workers at Windscale (NRPB R54) – (BNFL 119) mentioned in paragraph 10.47 above.

10.134 The criticisms of FRL are unjustified. In paragraph 232 of the Sixth Report the Royal Commission paid tribute to the range and thoroughness of the work undertaken by this organisation. I endorse that tribute. In doing so I do not wish to indicate that they are perfect. I have already referred to the lateness of their annual reports. In addition, as would be the case with any organisation, it was possible by way of an examination in detail of their activities over a period to point to particular matters in which their judgment may have been faulty. With the exception however of the lateness of their reports and their failure to take action more promptly when caesium discharges increased I was not satisfied that any of the charges made against them were justified. Neither of these two matters however affords any ground for saying that permission should be refused. I was, like the Royal Commission impressed with the work done and, in view of certain observations of Dr Bowen, I should say that it was plain that such work has not been limited to ascertaining whether any harm has been done in the previous twelve months but is also concerned with the prevention of harm in the future.

10.135 Before turning to NRPB there is one further matter which needs to be mentioned in connection with FRL and that is the question whether they publish sufficient of the information which they assemble. It is clear that they cannot publish everything and that it would serve no useful purpose if they did. On the other hand if they have, for example, made a large number of measurements which have established that certain possible pathways back to man are not critical, some short statement to this effect might be of value in two ways. Firstly, it would save anxieties as to whether such pathways had been considered. Secondly, it might enable a concerned and responsible person to direct attention to some other pathway which had not been considered. After the experience of their representatives under cross-examination at the Inquiry, I am confident that this aspect will be present to the minds of those in charge of FRL. It was plain that when they came to give evidence they did not appreciate the extent to which they might be required to give an account of their stewardship. The result was that suggestions made about inadequacies of their monitoring and research operations in the past were only dealt with at later stages in the Inquiry when material showing the extent of those operations was produced.

10.136 Dr Dolphin for NRPB admitted a mistake in methodology in the paper to which I have referred. I cannot regard this matter as indicating that NRPB are incompetent. Were it to be the case that a mistake of such a nature rendered the person concerned and the body to which he belonged incompetent, we would, if there were complete disclosure of all facts, probably have to regard all experts and all bodies to which they belong as incompetent. Whilst there are undoubtedly very many people whose mistakes are never discovered there are few if any who never in fact make mistakes.

10.137 There remains NII. I mention in paragraph 11.24 below a matter which should be given attention in the review of their activities presently being carried on. I have nothing to add.

10.138 On the matter of integrity and capability generally I see no ground for refusal of permission.

Miscellaneous

10.139 I have not given the question of BNFL's disposal site at Drigg more than a passing mention. This is because I was unable to see any significant risk likely to arise from it. This does not however mean that I regard the continued use of Drigg as something which can be left to look after itself. As with any site upon which radioactive substances are accumulated continuous vigilance is necessary. Escapes to groundwater and fires are instances of possible danger but such risks as there are, at present levels of usage and at such increased levels as are inherent if THORP is built, are minor compared with those risks which I have already considered. Since I regard those other risks themselves as being so small that, if properly understood, they would be acceptable, it follows that I regard the risks from the use of Drigg as acceptable also.

10.140 Lest there should be any misunderstanding I should also make it clear that the suggestions which I have made for certain improvements in the control system should not be taken to indicate that I regard the present system as inadequate to protect against emissions from THORP if built. I regard the system as adequate for this purpose but the searching examination to which it was subjected at the Inquiry revealed, as would such an examination of any institution, scope for improvement. I have therefore thought it right to draw attention to such matters.

11 Risks - Accidents

General

11.1 Accidents may conveniently be divided into:
  a. accidents involving a release of radioactivity which does not escape beyond the site boundary and thus does not expose the public;
  b. accidents which do involve a release beyond the site boundary;
  c. accidents during transport.

11.2 I shall consider each head separately but, before doing so, it is necessary to make some general observations. The first, and perhaps the most important, is that reprocessing is of such a nature that there is no danger that I can see of an accidental atomic explosion and no objector sought to suggest that there was any such danger. Since there were paraded before me a range of possibilities including even the shelling of the site in a civil war and aircraft crashing on to the site, I am confident that the risk of such an explosion occurring does not exist.

11.3 Secondly, whilst I accept that even the remotest of possibilities can happen, I reject completely the suggestion, which was made at the Inquiry, that, if something is possible it is also inevitable. It is a very remote possibility that a particular accident will occur if its occurrence requires, for example, the failure, not only of a primary source of supply such as electricity or water, but also the failure of one or more alternative sources of supply, and one or more automatic control systems together with more than one independent act of negligence. Improbable as it may be, however, it can happen. It is not however inevitable that it will happen. If it were, everyone, who, for example, drives his car tomorrow would be dead by tomorrow evening because it may truly be said of each of them that it is possible that he will be involved in a fatal accident during the day.

11.4 Thirdly, no real assessment of risks can be made when, as is the case with THORP, the design of a project is still at the conceptual stage. It will be for BNFL as designers and operators to develop and conduct the various safety disciplines as the design progresses and for NII to ensure that the plant does not operate until they are satisfied that it is safe.

11.5 Fourthly, I feel it important to draw attention to a point which causes much confusion. Reference is frequently made to the long periods of time during which a radioactive substance will remain radioactive and thus harmful. Whilst this is true it must not be forgotten that other substances which are not radioactive but which are harmful last, not merely for very long periods, but forever. It is only necessary to mention lead and mercury, both of which are released daily in considerable quantities from a variety of sources including fossil fuel power stations and, in the case of lead, exhaust from cars. Indeed Mr Patterson of FOE expressed himself as being more concerned about lead from exhaust than radioactive emissions from Windscale.

11.6 Lastly, it is necessary to mention the anxiety expressed by some about radioactivity being carried outside the perimeter by dogs, cats, rats, birds and so on. This is something which can without doubt occur but it must be remembered that dose limits are based on continuous exposure. It is, in my view, extremely unlikely that a bird or beast would be so heavily contaminated as to cause those with whom it had contact to be seriously exposed.

11.7 I have mentioned the NII. The Royal Commission recommended that the criteria and methods of working of the NII should be reviewed (para 531 conclusion 18). This recommendation was specifically made in relation to reactor safety but I have no doubt that it was intended to apply to nuclear installations generally and thus to Windscale and THORP. This recommendation was accepted in the White Paper (Annex A para 14). As in the case of risks from routine discharges it is not possible or desirable for me to attempt to make any findings about the safety of THORP as such. I can be concerned only with the possibility of building and operating it to acceptable standards of safety and with the machinery for ensuring that it is built and operated to such standards.

11.8 With regard to transport, the design and construction of containers are subject to IAEA regulations and the responsibility in this country is that of the DTp. Again I am concerned only with the possibility of transporting spent fuel to, and plutonium or fresh fuel from, Windscale at tolerable levels of risk, and with the machinery for ensuring that it is so transported. (Accident risks during transport are considered at paras 11.25 to 11.28 below.)

Risks not involving releases beyond perimeter

11.9 Such risks are to the workforce at the site. I have already (para 10.34) drawn attention to the fact that Mr Adams representing the workforce is in favour of the proposals and I see no reason to doubt BNFL's ability to operate to tolerable levels of risk. On this aspect of the matter the chief attack was made by WA through their witness Dr Wakstein, who sought to show that as a result of BNFL's past record they could not be relied on to operate safely. WA considered that before permission was granted there should be an independent investigation into the full details of all past incidents.

11.10 The information available on past accidents was limited. The one major incident was a 'blow-back' which occurred in B204 in 1973. This was the subject of a full report (Cmnd 5703) (BNFL 86). It involved the contaminatoin of 35 workers by ruthenium 106. In addition BNFL provided a list of all incidents which had been regarded as sufficiently serious to warrant a formal investigation and which had occurred in connection with reprocessing from its inception in 1950 to 1976. In this 26-year period there had been 177 such incidents. The list covered incidents involving contamination of workers and others not involving contamination. It gave however no details. Further information with regard to some of these incidents was later provided, but even this was of an abbreviated nature. The information was nevertheless sufficient to show that formal investigation is made of even the most innocuous incidents.

11.11 Having studied the information available and heard evidence from BNFL's and other witnesses I am satisfied, like Mr Adams, that BNFL are very safety conscious. I am also satisfied that THORP can be built and operated to tolerable levels of safety so far as on site risks are concerned. Nevertheless, an examination of such details as were available on the incidents discloses that many of them were due to comparatively simple errors in design, operating instructions or information, or in carrying out operating instructions. Errors of this sort can be found in every plant but I formed the distinct impression that more could have been done in the past and should be done in the future to ensure that procedures are sufficient for all eventualities, are strictly observed and continually rehearsed. In the absence of rehearsal there is an ever present risk that, by the time the particular eventuality occurs, the procedure which should be put into operation to deal with it will have been forgotten. BNFL were charged with complacency. This charge I reject. I find it quite unjustified. But it is beyond doubt that the nature of the risks is such that constant self reminders of their existence is necessary. A worker who, for example, has contamination on his hands will, if he fails to go through decontamination procedures probably feel no ill effects, yet he can expose both himself and others to harm.

Accidents involving releases beyond the site

11.12 Accidents going beyond the site may either be comparatively minor or of a major type such as to lead to severe contamination and death or injury beyond the site. No major accident of this sort has occurred in connection with reprocessing at Windscale. That releases of a minor nature may occur is plain, but given that BNFL and the control authorities exercise their present vigilance they do not represent a significant hazard. I have already mentioned the iodine release of 1972 in the previous section, but it may also be regarded as a minor accidental release as may the slight contamination of the beaches by tritium which has occurred on more than one occasion. Such releases are most unlikely to have caused any exposure beyond permitted limits. I need say no more concerning releases of this nature but I must deal further with the possibility of major releases.

11.13 There are, in essence, three possible sources for major releases, namely cooling ponds, HAWs, and the plutonium store. For security reasons no useful investigation could be made with regard to the last but I see no reason to doubt BNFL's evidence that any release of plutonium is less likely, and likely to be less harmful, than a release from HAWs. The possibility of such a release was investigated in some detail.

11.14 The HAWs contain the largest concentration of radioactivity. Their vulnerability lies, if it lies anywhere, in the fact that the liquid has to be kept cool and the cooling process involves both water supply and electricity. Mr J. K. Donoghue, Manager of the Safety Assessment Group in the reprocessing division at Windscale, had considered the theoretical possibility of a loss of cooling in a tank leading to boiling. He had carried out experiments which led to the conclusion that the resulting discharge to the atmosphere would at most amount to one part in 10,000 of the total radioactivity in the tank, that is a release of 10,000 curies. The Safety and Reliability Directorate (SRD) of the UKAEA had written a computer programme to determine the consequences of releases to atmosphere of radioactive material. Assuming the release postulated by Mr Donoghue, SRD had, by the use of the programme, calculated that, in the most unfavourable circumstances envisaged, it might lead to 10 persons contracting cancer, the necessity to evacuate for a few days persons living within one mile of the works and the ban for a few weeks on consumption of foodstuffs produced within a 10 mile radius. Two principal questions which arise out of the foregoing are (a) how likely is it that such a release would occur and (b) should some worse occurrence be regarded as realistic albeit very improbable.

11.15 Mr Donoghue stated in evidence that the chance of the loss of cooling necessary to produce the postulated release had been estimated as being 1 in 1 million years. His estimate in his written proof was 1 in 100,000 years but he corrected this. I do not know how accurate this is, nor do I know how the estimate was arrived at but it was established in evidence that each tank would be provided with at least two spare cooling circuits, and that there were four separate available sources of water and three separate sources of electricity. In addition, if all these failed for some reason or another, it would take a number of hours for the contents to reach boiling point and days before boil-away was complete, an event which would be necessary before Mr Donoghue's supposed release could be achieved. The number of hours and days respectively depends on the assumptions made but I was provided with calculations on a series of different assumptions. These showed times to boiling point, ranging from about 9 hours to 31 hours and times to boil-away ranging from about $2\frac{1}{2}$ days to 8 days. To obtain the release contemplated would therefore require not only such a series of failures as to render the probability very remote but also an absence of remedial action for a minimum of $2\frac{1}{2}$ days, a situation which appears to be possible only in the event of some disaster of a severity which would render the consequences of the postulated release itself of minor concern.

11.16 Mr Donoghue stated that his colleagues at BNFL regarded his contemplated release as impossible. Having heard all the evidence I am not surprised, nor am I surprised that Dr G. R. Thompson, who was the principal objector's witness on this subject, expressed the view that he considered it likely to be possible to build THORP to acceptably safe limits.

11.17 In the light of the above I find it unnecessary to consider a series of other releases which were suggested, all of which were necessarily regarded as being less likely. They involved such things as inattention and lack of remedial action for much longer periods, loss of pond water in the fuel element storage ponds with similar inattention, aircraft crashed on to or into the sides of HAWs and the like. Dr Thompson accepted that the various exercises which he had carried out were not intended to be predictions of risk under realistic conditions.

11.18 When considering the risks involved it is also necessary to consider the position if there is no reprocessing. In that event the spend fuel with its entire content of radioactivity including the plutonium would have to be stored. I see no reason to suppose that this would represent any lesser hazard. Indeed there was some evidence that it might represent a greater one. Plutonium separated and carefully stored might well be less likely to suffer accidental release than if left in fuel rods. Storage facilities for fuel rods might for this reason alone be a greater hazard than HAWs. In the end I reach the firm conclusions on the evidence, first, that THORP can be built to tolerable levels of safety and secondly that not to reprocess is unlikely to present any lesser hazard.

Accident risks – Industrial action

11.19 Before leaving this subject it is necessary to mention the question of hazard through industrial action. One objector (objecting as regards reprocessing of foreign fuel only) expressed strong views that the safety of the public demanded that the workforce at Windscale should give up, and be proud to give up, the right to strike, in exchange for terms and conditions guaranteed equal to the best in equivalent posts in industry generally. Any disputes should, he thought, be dealt with by arbitration and the artibtration award should be binding. He considered that the workforce would thus become an elite workforce.

11.20 The objector concerned was a Mr W. C. Robertson. He had himself worked at Windscale, is a fellow of the Institution of Electrical Engineers and had had very considerable experience in electrical and mechanical engineering. Bearing in mind that published newspaper reports of the strike, which had occurred at Windscale early in 1977, had suggested that the point of danger to the public had been very closely approached as a result of the strike, it appeared to me necessary, not only to explore the possible consequences of a strike but also to seek the views of Mr Adams (as representing local workers), the TUC and BNFL on the suggestions made by Mr Robertson.

11.21 BNFL, the TUC and Mr Adams were at one in regarding restriction on the right to strike as being unnecessary. All took the view that whilst there would have to be agreed procedures to ensure that a strike would not endanger the public such procedures would be sufficient. No doubt this would be true if it were possible to ensure (a) that there could be no disagreement about the point at which public safety became overriding and (b) that the workforce on strike would always comply with the procedure agreed. Neither is possible. It is commonplace to find that there is disagreement about the imminence of danger and equally commonplace to find that men on strike do not follow their leaders' advice or directions.

11.22 If, therefore, it had appeared to me that the absence of the workforce or a part of it would be likely to create significant hazards I should have had no hesitation in endorsing Mr Robertson's suggestions. Having investigated the matter I do not consider that it is likely provided that picketing in aid of a strike does not prevent either:
a. delivery of essential supplies or
b. the attendance without hindrance of a small safety force to maintain surveillance and take any remedial action necessary in the event of e.g. the failure of one source of electric or water supply.

Even if it does the public can no doubt be protected, as was pointed out in the 1977 strike, by the use of troops unless the situation is such that special expertise is required from them. This does not, however, provide a particularly reassuring answer to the public. There is always a reluctance to employ troops until their employment is absolutely unavoidable. This reluctance is due to the escalation of the dispute which is likely to follow upon the use of troops. This being so there is necessarily a possibility that there may be a misjudgment as to the moment when their use can be held off no longer.

11.23 This matter is not peculiar to the THORP facility on the site, or to Windscale, or indeed to the nuclear industry. It is not appropriate that I should make any recommendation. Having, however, heard evidence from a number of local witnesses I have no doubt that the local public would be greatly relieved if they knew for certain that no matter what industrial action was taking place there would be no hindrance to the delivery of essential supplies or to the attendance of a safety staff on site.

Adequacy of the system

11.24 It was suggested that NII were not sufficiently involved at all stages of design to ensure ultimate safety and that they were in any event not equipped with sufficient scientific expertise to check the designs. The former suggestion I reject. Mr H. J. Dunster, the Deputy Director General of the Health and Safety Executive, gave evidence on this point which I found convincing. For the Inspectorate to become more intimately involved has two disadvantages. In the first place it may lead to the designers seeking to shed what is properly their responsibility. In the second place an Inspectorate which is too intimately involved can be robbed to an extent of the independent objective judgment which is its function to exercise. The second matter is one to which attention should be given. I make no finding that NII are inadequately equipped but it is a matter which should certainly receive attention in the review which is presently taking place. Their task is to pass judgment on plants which are designed by very highly qualified experts and they must, if they are to perform their function, have, or have access to, at least equal expertise. It was not established to my satisfaction that this is the position.

Risks during transport

11.25 Whether or not reprocessing takes place at Windscale, spent fuel will be transported there from reactor sites or, in the case of foreign spent fuel, from ports of arrival. Such transport will involve no new risks. It will take place under stringent regulations and in massive flasks. I am satisfied on the evidence that the transport of spent fuel creates no significant risk and that such risks as may exist are less than those involved in the transport of other substances which cause no alarm to any substantial section of the public. The position appeared to me to be so clear that I say no more about it.

11.26 After reprocessing there are the following forms of transport to be considered:
 a. Transport of fresh fuel to reactor sites or ports, possibly irradiated as a protection against theft.
 b. Transport of plutonium to the foreign suppliers of spent fuel.
 c. Transport of vitrified highly active waste.

11.27 The first of these would present no greater risk than transport of the spent fuel to Windscale. As to the second I have already mentioned that plutonium has been transported from Windscale in the past. It has been so transported in containers which have been tested by being dropped 2,000 feet on to concrete by aircraft. I can see no significant escalation of risk in this connection. With regard to transport of vitrified highly active waste none has yet taken place. I was, however, given no reason to suppose that risks involved could not be kept to tolerable levels.

11.28 One matter which was discussed at some length was the possibility of a ship containing spent or fresh fuel or plutonium sinking, perhaps in conjunction with a boiler room fire or explosion. Here again risks appear to me to be slight. If a ship were to sink in a depth in which salvage operations were possible the containers would survive more than long enough for salvage to take place. Risks from fire or explosion could, it appears to me, be rendered very remote by relatively simple protective measures and, even if the worst were to happen, the amount of radioactivity contained in any one consignment would, if it were released in the ocean, be unlikely to cause significant harm. This would also be the case if a ship were sunk in waters in which salvage was not possible.

The emergency plan and local liaison committee

11.29 In connection with accidents it is necessary to mention the Windscale local liaison committee which was established in 1957 (following a reactor pile fire) in order to draw up a plan to be operated in the event of a serious incident which affected members of the public. I refer to this plan simply as the emergency plan.

11.30 It should not be necessary to say it but an emergency plan is of little or no use if those who have to act upon it do not know what it is until an emergency which calls for it to be put into effect has arisen. The reason why I mention this very obvious fact is because it emerged in evidence that some of those who, in the event of an emergency, would be required to take action under the plan which was formulated did not even know they had any responsibilities, much less what those responsibilities were. This was clearly a grave defect as was acknowledged both by Cumbria and BNFL. The matter first arose on Day 60 when Nurse Florence Corkhill (retired district nurse) was giving evidence. Thereafter I was informed that improvements were being undertaken. It is essential that this should happen, albeit that it may be, and in my view is, unlikely that the plan will need to be put into operation. I make no specific recommendations. I observe merely that it is a matter which involves close co-operation between the various authorities concerned, and regular checks to ensure both that all who have responsibilities are aware of them and that the materials necessary for carrying out such responsibilities are available and in good condition.

11.31 The Government have accepted (White Paper para 31) the Royal Commission's recommendation that in principle all emergency plans for all civil nuclear installations should be made available to the public. This marks a change in policy and should remove any inhibition previously felt about disclosing details. Indeed during the course of the Inquiry both BNFL and Cumbria made their plans public by putting them in as evidence (BNFL 306 and CCC 31). As part of the consideration now being given to the steps which can and should be taken as a result of the change in policy, BNFL, Cumbria and others concerned will no doubt give attention to the question whether certain members of the public should be given specific information as to action to be taken in certain events, e.g. local farmers might be informed that in certain circumstances, possibly indicated by some audible signal, supplies of milk should be held up pending testing.

11.32 The Local Liaison Committee comprises representatives of BNFL, Cumbria, Copeland, the Cumbria Health Authority, the authorising departments, the police and the appropriate trade unions. It has met some 30 times. All meetings have been held at Windscale under the chairmanship of a BNFL representative.
Its terms of reference were:
'To define responsibilities and action to be taken by all interested parties in the event of a district hazard arising from a site incident at Windscale and Calder Works, to re-assure local opinion of the hazards involved and to create an administrative machinery for the protection of the population in the event of a serious incident.'

11.33 The minutes for the 14 March 1975 record however a rather broader approach. The Chairman then suggested that it might conduct its business under the following heads:
 'a. Future developments and their potential impact on the community and the environment prior to construction.
  b. Results of analysis of potential environmental impact prior to operation.
  c. Experience at the Works.
  d. Results of environmental monitoring.
  e. Review of emergency schemes.'
The committee considered that this was a helpful formal rationalisation of what had in fact been past practice.

11.34 As a vehicle for keeping the public informed or reassuring the public the committee has plainly failed to carry out its terms of reference. Local witnesses, even in one case a local councillor, had not even heard of its existence much less its activities. It was generally accepted that the committee was in need of reorganisation and that its functions required redefinition. They plainly do. Such reorganisation and definition of functions are matters for agreement between those concerned. The weaknesses of the present situation having been exposed at the Inquiry I have no doubt that these matters will be given urgent consideration and that agreement will be reached. The most helpful action that I can take is to suggest certain matters which appeared to me on the evidence to be desirable. These are:
  i. The chairman should not automatically be a representative of BNFL. He should be elected by the membership.
  ii. Meetings should not normally be held at Windscale, and it is desirable that a summary of the proceedings at each meeting should be made public. Public attendance should also be considered.
  iii. The committee members or selected representatives should periodically visit both the works and the disposal site at Drigg.
  iv. The committee should not try to take on too many tasks and might well consider allotting specific tasks to sub-committees. To ensure representation of all interetsed parties membership will necessarily be fairly numerous – too numerous for the efficient conduct of business.

11.35 I suggest the foregoing matters because it was clear from the evidence that such efforts as had been made in the past to inform the public and save it from needless anxieties had signally failed. In this connection BNFL have accepted that it should be a planning condition, if permission is given, that they should provide Cumbria with the results of environmental monitoring and with reports of all incidents at the works which are reportable to the Secretary of State for Energy. Armed with this information, and no doubt advised as to its implications by Professor Fremlin or some other well qualified independent expert, Cumbria would consider publication of periodical reports, in some digestible form. Such reports whether by Cumbria or by a reformed Liasion Committee are clearly desirable.

12 Size of Plant

12.1 Dr Barry Shorthouse, a member of the Open University's staff and Research Director of the Higher Educational Management Institute, giving evidence on behalf of WA, was of the opinion that the construction of a 1200 tonne per annum plant represented too great a jump in size having regard to the facts:
1. that neither BNFL nor anyone else had successfully operated oxide fuel reprocessing on a regular production basis;
2. that BNFL intended to proceed to the full size plant from
    i. a 1/5000 scale pilot plant using fully active material which would replicate the THORP process from the stage of dissolving of the fuel to the purification of the uranium and plutonium nitrate products by solvent extraction;
    ii. a full scale test rig, comprising the pulsed columns, mixer settlers and associated equipment proposed for THORP, operating with uranium solutions and not with highly active material.

12.2 It emerged in his evidence that Dr Shorthouse's concern mainly related to the use of pulsed columns as opposed to mixer settlers for the separation of the uranium and plutonium from the fission products. He took the view that it would only be prudent to proceed from the fully-active miniature pilot plant to a plant 10 times its size and that the stage by stage process could not be avoided by the use of the full size test rig using uranium solutions only, since such a rig would not show the effect of radiation on the solvent or of a possible modification by suspended particles of the droplets produced in the perforated plates of the pulsed columns.

12.3 That scaling up can reveal problems which were not apparent in small scale pilot plants is clear. Mr B. F. Warner was fully aware of this as indeed is any designer. Nevertheless, Mr Warner has 28 years of experience dating from the design of the original plutonium separation plant. He was confident that from the fully active pilot plant, the full scale rig and information to which he had access concerning the use of pulsed columns, he could obtain the necessary material for a final design of THORP to reliable and safe working standards.

12.4 Mr Warner showed himself to be not only a very impressive witness but one who approached the matter of design with great caution and he was ready to draw attention himself to points of uncertainty, which had not been observed by objectors. I accept his evidence that to proceed direct to THORP by the route which he proposed is prudent; the more particularly when NII will be there to check the final design and operation before, if permission is granted, THORP can go into production.

13 Public Hostility

13.1 That there exists in some proportion of the public a degree of hostility to, or anxiety about, nuclear power, and thus reprocessing, is beyond all doubt. What is the strength of the hostility or anxiety and in what proportion of the public it exists are however matters which I was unable to assess with any accuracy. I doubt whether such matters are in fact capable of any assessment for the hositility and anxiety stem from a great variety of matters and vary in strength from person to person. In some cases anxiety and hostility can be dispelled by greater knowledge, in others they may be increased, whilst in yet others they will remain no matter that those who feel them recognise them to be irrational. Nevertheless a number of matters of some importance emerged during the Inquiry to which it is desirable that I should draw attention.

13.2 In the first place, much of the anxiety is caused, or is increased, by the facts (i) that many films or books about peaceful nuclear power are prefaced by or include sequences showing, or containing illustrations of, the mushroom cloud of an atomic explosion, and contain explicit or implicit reminders of the results of Hiroshima and Nagasaki, (ii) that some films or books contain emotive and inaccurate pictures or statements, and (iii) that firm assurances of safety are sometimes given which do not reveal the true position and, as a consequence, create suspicion when the true position is revealed. I will give an example of both (ii) and (iii) above. No examples are needed in respect of (i).

13.3 On behalf of WA I was shown a film entitled 'Caging the Dragon' made by their witness Dr C Wakstein, an engineer and film maker. It dealt to a large extent with the question of safety at Windscale. At the beginning of the film there was a shot of Windscale closely followed by pictures of a man who had suffered severe radiation burns. Shortly thereafter there was a reference to a fire which had occurred in 1957 in one of the original reactors at Windscale followed by a sequence showing huge flames emerging from the top of a tower. Pictures of these flames were repeated twice more during the course of the film. The copy of the film shown to me suffered from bad synchronisation but I formed the distinct impression that the pictures of the flames were intended to be taken as, and would be understood to be, actual pictures of what had occurred at Windscale and that the pictures of the radiation victim were intended to be taken as, and would be understood to be, pictures of a man who had in fact suffered his injuries at Windscale. I subsequently asked Dr Wakstein about both matters. He informed me that the pictures of the flames were in fact pictures of the flames from a flare stack at a coke works in Sheffield and that pictures of the radiation victim had been taken from a medical journal and represented a victim of an accident in connection with the weapons industry in the USA. He accepted that someone looking at the film might have thought the fire did show what had happened at Windscale in 1957. He also accepted that the suggestion was that the victim had suffered his burns in the civil nuclear industry albeit not at Windscale. The photographs of the victim were however intended merely to show what he considered the potential risks in the civil nuclear industry.

13.4 I considered both these matters to be seriously misleading and particularly undesirable in a film made and shown by a professional engineer such as Dr Wakstein. It was later submitted that my reaction to the film had been over-critical and did not represent what an ordinary person looking at the film and listening to the soundtrack could infer (WA232). Having only seen the film once and with poor synchronisation I made arrangements to see it again whilst writing this report. I saw it, this time with good synchronisation, on 18 January. Having done so my initial impression as to the flames is strengthened. I have no doubt that the ordinary viewer would understand the sequences to be actual pictures of what had happened at Windscale and that the repetition of the pictures would have been taken as a reminder that the very serious fire depicted had already occurred there and might easily occur again. As to the pictures of the radiation victim, I remain of the opinion that many viewers would take the pictures to represent the victim of a Windscale accident and that almost all would understand them to represent an actual occurrence in normal commercial nuclear operations. Both impressions would be equally false. If a film containing such sequences is to be used the viewing public should at least be given an account of the true facts but I doubt whether even this would be satisfactory.* The visual impression is so strong that it might not be removed by an explanation.

13.5 With regard to assurances of safety, Professor Ellis referred to a statement in paragraph 3 of the Background

---

*I have been informed that Dr Wakstein does intend in the future to explain the position when the film is shown. I report this by way of footnote for in fairness to him it is right that it should be known that he has taken this decision.

Note issued by the DOE, MAFF and DTp (G.1) and put in at the Inquiry that 'the Authorising Departments are satisfied that the radiation doses to the most highly exposed members of the public have generally been no more than a small fraction of the ICRP limits.' The FRL reports for 1975 and 1976 show the maximum consumer of fish in the local fishing community as receiving 34 per cent ICRP in 1975 and 44 per cent in 1976. Professor Ellis did not regard 34 per cent as being a small fraction. Nor do I. It is no doubt true that 'in general', as the quotation says, percentages were only small fractions of ICRP limits, but I have no doubt that it would have given a truer picture had the statement said that, owing to caesium discharges, the percentage reached by maximum fish eaters had risen rapidly to substantial fractions of ICRP limits in the last three years but that this situation would be very greatly improved when a treatment plant for which planning permission had already been granted came into operation.

13.6 The second matter to which I draw attention is that much of the opposition appeared to be based on sincerely held moral grounds, and that amongst those who advanced opposition on such grounds there was a tendency to suggest that supporters were acting in an immoral way. This attitude is plainly unsustainable for it is clearly possible to hold an equally sincere belief that reprocessing is necessary on moral grounds. If by reprocessing we can lessen the amount of plutonium to the escape of which future generations may be exposed; if we can avoid the possibly greater harm to both present and future generations which would result from deriving energy from the mining and burning of coal instead of from nuclear power; if we can, by using nuclear power, save society in 20 years' time from the troubles that would follow upon a reduction in living standards; if at the same time even a modest number of the unemployed can obtain employment, and if this can be achieved at the cost of an insignificant exposure of ourselves to radiation, it may be that support is the moral answer. It is not for me to attempt to reach a conclusion on the morality of the situation. It can however be stated that, as was accepted by Dr Spearing, the question is not as easy as some believe or would wish to believe. It was, moreover, abundantly clear on the evidence that some who pursued the moral line had done so without investigating the consequences of pursuing alternatives.

13.7 Thirdly it is important to observe that support for nuclear power and even reprocessing does not come only from those within the nuclear industry. Cumbria, advised by Professor Fremlin, Copeland, the relevant House of Commons Select Committee on Science and Technology, the EEC Commission, Sir Brian Flowers (BNFL 236), Dr Chapman for FOE, Dr Bowen for IOM and Professor Radford for NNC, all favoured the continued growth of nuclear power.

13.8 There were submissions as to the dire consequences of granting permission but little or no evidence to support those submissions. I am satisfied on the evidence that although hostility does exist its existence is not so widespread as to justify a refusal of permission on that ground alone.

13.9 I conclude this section by referring to a matter which concerned a number of objectors namely that the support of the local councils should not be given too much weight, since their decision to support BNFL's application was arrived at on only a fraction of the information which emerged at the Inquiry. I agree with the point made, but I observe that it is a point which applies with equal or more force to objectors and particularly to members of the public who may have been opposed to the application. Cumbria, before arriving at their decision consulted, in addition to Professor Fremlin, a number of bodies concerned with the interests of the public. Members of the public will, for the most part, have had much less information than had Cumbria. If, therefore, Cumbria's and Copeland's decisions to support should receive little weight, so also should the views of those who were at the start of the Inquiry opposed to the application. Partly for this reason and partly because the questions put were such as to render the results valueless, no weight at all can be given to the results of a number of opinion polls or petitions to which I was referred.

14 Conventional Planning Issues

General

14.1 Conventional planning issues, by which I mean, broadly, those matters which are mentioned in points 3–6 inclusive of the Rule 6(1) Statement (see paragraph 1.2 above) do not arise unless the first of the three questions posed in paragraph 1.7 above is answered in the affirmative, i.e. unless it is the case that oxide fuel from UK reactors should be reprocessed in this country. For the purposes of this section I shall therefore assume that this is so.

14.2 Before dealing specifically with the various matters which arise it is necessary to mention generally the question of the implications of THORP for local employment – point 5 of the Rule 6(1) statement. Since unemployment over the last year in the areas of West Cumbria most likely to be affected by the proposed development has ranged between 7.4 per cent and 12.3 per cent of the working population, and since unemployment levels in the area have for many years been higher than the national average, this is a matter which is of great concern to local people. BNFL did not make it any central part of their case that the proposed development would make any very significant impact upon the unemployment problem. They did however contend that it would have some beneficial effect on the problem and that it would, in addition, provide some measure of economic growth and some migration into an area suffering from the ill effects of emigration. Objectors were nevertheless desirous of establishing that the beneficial effect, if any, would be very small, and that there might indeed be some adverse effect upon the unemployment problem. They believed that local support for, or failure to object to, the proposed development was based in large measure upon the mistaken belief that THORP would significantly alleviate the unemployment problem.

14.3 It appeared to me at a comparatively early stage, that, although BNFL were probably right in contending that there would be some measure of local benefit in the three respects mentioned, it could not be great enough to justify any requirement that the workforce, the people of Cumbria, the public at large or future generations should accept any exceptional risk. This was accepted by the applicants and their supporters and, this being so, it appeared to me that it would be an unjustifiable waste of time and money to conduct an exhaustive examination into what was likely to happen in relation to employment during and after a 10 year period. I therefore urged all parties to try to agree on a range of forecasts within which the employment effects of THORP would be likely to fall. Although this proved impossible, agreement was reached on a large number of facts. As a result the time occupied on this problem was considerably shortened and I shall consider the matter more briefly hereafter than would otherwise have been necessary.

14.4 I deal next with the various matters which arise, doing so as far as possible under the headings used in the Rule 6(1) Statement. But before going to the first of these headings it is necessary to consider whether, on the assumption made, Windscale is a suitable site. In considering this matter I shall also assume that the proposed development consists in the 1,200 tonne p.a. plant proposed.

Suitability of the site

14.5 The principal matters urged in favour of Windscale as the site for such a plant were as follows:
  i. The information acquired over a period of some 25 years in relation to discharges of radioactive material from Windscale would be of great value in protecting the public from any harmful effects from the discharges from THORP.
  ii. The existence of reprocessing and plutonium handling and storage facilities at Windscale made the further development of the site preferable to the creation of new facilities elsewhere.
  iii. The expertise for reprocessing was concentrated in the staff already at Windscale.
  iv. Although it would probably be necessary to find another site if a further reprocessing plant was subsequently required, to multiply sites before it was necessary was undesirable, for it would multiply the places at which plutonium had to be protected and the routes over which spent fuel and, in the future, plutonium and fresh fuel had to be carried.
  v. Since all fuel containing plutonium would be manufactured at Windscale, reprocessing facilities there would minimise the transport of plutonium. The suggestion that reprocessing facilities, if required, should be at all reactor sites would multiply transport movements since, unless reactor sites also had fuel fabrication facilities, they would have to send their plutonium to Windscale.
  vi. To have reprocessing facilities at reactor sites

would be economically unsound, for each plant would be small and the economies of scale would therefore be lost.

14.6 Against Windscale as a site it was principally contended that:
  i. The site was inherently unsuitable from the point of view of health risks, for discharges were into the semi-enclosed Irish Sea with much sandy or muddy coastline relatively close to the discharge point.
  ii. It would be preferable to site such a plant on a rocky coastline of the open ocean, with current patterns which would carry the discharges away from the coasts.
  iii. Radiation exposure of the public would be greater if the plant were at Windscale than if it were elsewhere, since THORP discharges would add to exposures from present and future magnox reprocessing.
  iv. It was undesirable to site such a plant next to an area of Great Landscape or Scientific Value and close to a National Park.
  v. The proposed development would involve a fundamental departure from the County Development Plan.

14.7 More generally it was urged on behalf of objectors that no permission should be given until an environmental impact analysis (EIA) had been made and possible alternative sites had been considered and shown to be unsatisfactory.

14.8 All of the points urged in favour of Windscale were established to my satisfaction as valid points both of fact and argument. Collectively they have considerable weight. Of the specific matters urged against
  a. I accept that points (i) and (ii) would be valid points if the Windscale works did not exist and had not been functioning for many years. Both points were raised principally by IOM and their witness Dr Bowen. Dr Bowen, however, made it clear that, although he regarded the site, had it been virgin, as unsuitable, he was not contending that permission should be refused; only that there should be certain preconditions. He said in evidence:
      'I believe the store of information which has been gathered about the present reprocessing plant in that site is an enormous plus in argument for its continued use for this purpose in spite of its inherent unsuitability.
      Q. Because in projecting into the future the effects of the oxide reprocessing plant, one starts with, at any rate, a considerable body of knowledge as to the effect of what has happened in the past?
      A. Precisely, yes.'
      I accept Dr Bowen's view of this matter. I am satisfied that the value of the information available outweighs the inherent disadvantage.
  b. As to the third point, it is plainly the case that there will be additional exposure but, since the total exposure will, it is assumed for present purposes, be within tolerable levels the additional exposure involved is not of great significance.
  c. Point (iv), bearing in mind what is already at Windscale and the total levels of discharge assumed, appears to me to be of little weight. I would add that Dr G Halliday of the Cumbria Naturalists Trust, a body which studies and protects places of interest to naturalists and scientists, gave evidence that he had no reason to suppose that present discharges from Windscale were having an adverse effect on wild life.
  d. As to the last specific point the site falls within an area designated 'Large area for industrial use'. The proposed development would not, in my view, constitute a fundamental departure from the development plan but, even if this is wrong, the only effect would be to increase the burden of proof upon the applicants by requiring them to justify the departure. If such justification is required I am satisfied that, if there is to be a reprocessing plant at all, the points urged in favour of Windscale, after taking into account the points against, fully justify that departure.

14.9 With regard to the general points urged on behalf of objectors:
  i. There is no legal requirement for an EIA. I am satisfied that all matters which might or would have been included in an EIA were properly investigated at the Inquiry. This was largely accepted. It is possible but by no means certain that, had there been such an analysis, it would have saved time at the Inquiry and it is a matter for consideration whether in particular cases in the future an applicant or planning authority should be required to prepare such an analysis. All that it is necessary for me to say is that the absence of an EIA in this case affords, in my view, no ground for refusing permission.
  ii. There is no legal obligation to consider and rule out alternatives. It is plain that BNFL had not done so. They took the view that the case for Windscale was so clear that a consideration of alternatives was unnecessary. They were in my view correct in so thinking.

14.10 In the light of the foregoing I conclude without hesitation that if oxide fuel is going to be reprocessed up to an amount of 1,200 tonnes per annum Windscale is a suitable and proper site, indeed *the* proper site.

Effect on amenity

Visual impact

14.11 In assessing the likely visual impact of the proposed development, I had the benefit of a number of photographs and a large model of the site together with drawings of the profiles of the buildings. In addition to my visit to the site, I also travelled round the area of Windscale in order to see the present site from a wide variety of viewpoints. Evidence by the objectors on visual impact was given principally by Professor G Ashworth, Professor of Urban and Environmental Studies at the University of Salford, for the IOM, Mr C C Thirlwall, Planning and Transportation consultant for the TCPA and Mr Geoffrey Berry for the Friends of the Lake District (FLD).

14.12 The site at present comprises an untidy collection of buildings and plant and the part of the site on which it is intended that THORP and the associated buildings should be built is partly vacant and partly occupied by low industrial buildings. Development on that part of the site would in my view afford an opportunity for improving its appearance. This was accepted as a possibility by Professor Ashworth, from whose evidence I quote:

'Q. You have seen Windscale, I suppose, have you?
A. Yes, sir.
Q. Do you think it is at least possible that it might look very much better with some slightly more – I do not wish to tread on anybody's toes – but some slightly more pleasing buildings in the middle of it?
A. Well, I think that is difficult to judge on the basis of the information that we have been given so far. I might go so far as to agree with you, sir, in saying that it may be that it will be made no worse.
Q. It would be difficult, would not it?
A. To make it any worse?
Q. Yes.
A. Indeed, sir.'

14.13 It might nevertheless be possible that the addition of further large structures on the site would be unacceptably intrusive. This indeed was the case put forward by Mr Berry.

14.14 In support of his case he tabled a number of very good photographs and relied particularly on photograph FLD 5A to make the point that the proposed development would be intrusive. With Mr Berry's consent, BNFL reproduced the photograph again with the proposed THORP buildings drawn on it, together with the additional magnox facilities for which permission has already been given. This photograph FLD 5B does not, in my view, bear out Mr Berry's contention.

14.15 In addition to the case on this matter put forward by BNFL and the County Council, I also heard evidence from Mr R. B. Baynes that the Lake District Special Planning Board would have no objection to the proposed development on grounds of visual impact, given the size and nature of the existing buildings on the site.

14.16 This application is for outline permission only, and the detailed design and overall dimensions and precise layout of the structure have, as is common in such cases, not yet been settled. I am nevertheless satisfied on the evidence before me that there is no case for refusing permission on grounds of visual impact, and that, if permission is granted, there is scope for improving the appearance of the site.

Noise and nuisance

14.17 The only further substantial objection on amenity grounds was that the proposed development would involve substantial construction work over a number of years, which would cause noise and nuisance and which would generate additional traffic movements. It should however be noted that fairly extensive construction activities already take place at Windscale and have for many years past. Those activities will continue well into the future in any event in respect of the planning permissions recently granted for magnox and storage facilities. There was no evidence before me that the additional activities resulting from THORP would have an unacceptable effect on amenity and I therefore conclude that the effect of the proposed development on amenities affords no ground for refusing permission.

Implications for local employment

14.18 Although this matter is referred to after traffic in the Rule 6(1) statement it is convenient to consider it now since the effect upon traffic depends in part on the employment question. As indicated in paragraph 14.3 above I shall deal with this question briefly.

14.19 It is estimated that the proposed development would provide 1,000 additional permanent jobs, the increase taking place over the 10 year period from 1978–1987. The steps by which this increase is estimated to be achieved, broken down between industrial (skilled, semi-skilled and unskilled) workers and non-industrial (scientific and administrative) workers is shown in the table below.

Year	Industrial	Non-industrial	Total
1978	100	120	220
1979	—	—	—
1980	—	—	—
1981	—	—	—
1982	—	—	—
1983	50	10	60
1984	75	10	85
1985	120	15	135
1986	233	67	300
1987	167	33	200
	745	255	1,000

14.20 These additional jobs will be filled partly by those already residing in the area and partly by those who will be attracted into the area. In so far as they are locally filled they will be filled as to part from the unemployed

By courtesy of the UKAEA
British Nuclear Fuels Limited, Windscale Works

An aerial view of the Windscale and Calder Works, in Cumbria, of British Nuclear Fuels Limited. On the right is the Calder Hall nuclear power station. Opened in October 1956, it was the world's first large-scale nuclear power station.
Since this photograph was taken the course of the River Calder has been straightened.

By courtesy of Mr G Berry of 'Friends of the Lake District'
The works (Ref 5A/A/2).

By courtesy of British Nuclear Fuels Limited
As photo opposite with the proposed new THORP Plant and the new magnox facilities superimposed.

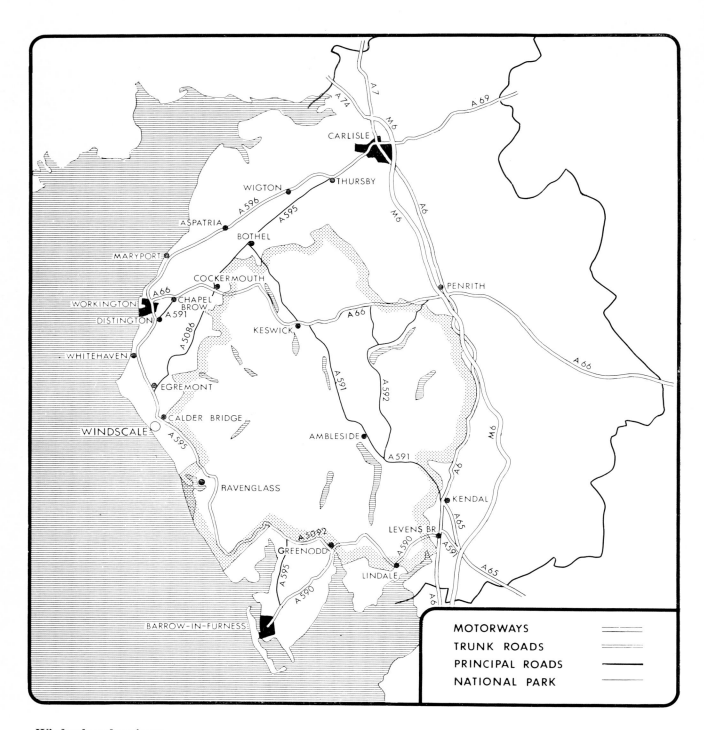

Windscale and environs

This map was not submitted as part of the Report but has been included at the request of the Inspector.

pool and in part by personnel transferring to Windscale from other jobs. In so far as they are filled by immigrants, some immigrants will bring wives and families who are also, or will later be, seeking employment in the area. On the other hand the extra jobs will increase the economic activity in the area, which will itself create further jobs.

14.21 What precisely will happen it is unnecessary to ascertain with any degree of accuracy. It is to be observed however that, if THORP were promptly permitted, the additional jobs in the first five years would total only 220. Even if as many as half this amount were filled from the locally unemployed the effect upon the unemployment problem in the area would thus be small.

14.22 The facts relating to the past are best set out in the form of a table.

14.23 It will be seen from the foregoing that BNFL's record in local recruitment over the past three years has been very good and they intend to recruit as many workers as possible locally. The figures available, however, do not cover a sufficiently long period,

BNFL Local Recruitment

Category and year	Net increase in workforce	Total number recruited	Total local recruitment	Local recruitment as % of total	Local recruitment from unemployed	Local recruitment from unemployed as % of local recruitment
All employees						
Year ending 31.3.75	204	509	434	86%	—	—
" " 31.3.76	518	837	584	70%	201	34%
" " 31.3.77	333	800	604	76%	193	32%
Total for three years (or average)	1,055	2,146	1,622	75% (av)	—	—
Craftsmen						
Year ending 31.3.75		120	104	87%	—	—
" " 31.3.76		173	96	55%	19	20%
" " 31.3.77		251	166	66%	23	14%
Total for three years (or average)		544	366	67% (av)	—	—
Non craft						
Year ending 31.3.75		205	203	99%	—	—
" " 31.3.76		369	335	91%	132	39%
" " 31.3.77		329	312	95%	98	31%
Total for three years (or average)		903	850	94% (av)	—	—
Professional						
Year ending 31.3.75		33	5	15%	—	—
" " 31.3.76		95	7	7%	4	57%
" " 31.3.77		38	6	16%	5	83%
Total for three years (or average)		166	18	11% (av)	—	—
Scientific						
Year ending 31.3.75		56	36	64%	—	—
" " 31.3.76		68	44	65%	18	41%
" " 31.3.77		50	31	62%	29	94%
Total for three years (or average)		174	111	64% (av)	—	—
Admin and others						
Year ending 31.3.75		95	86	90%	—	—
" " 31.3.76		132	102	77%	28	27%
" " 31.3.77		132	89	67%	38	43%
		359	277	77%	—	—

Notes:
1 Local Recruitment is recruitment from what BNFL describe as their local catchment area and extends from Workington to Millom.
2 Figures for Recruitment under the heading 'Craft' include Apprentices completing apprenticeships in the year and entering full employment but not those beginning apprenticeships.
3 The heading Recruitment from unemployed includes those employed from amongst school and university leavers.
(Refs CCC 34 Tables 1–10 and TCPA 97.)

nor, within the period covered, are they sufficiently comprehensive, to enable any reliable projections for the future to be based upon them. Cumbria produced an estimate that, of the total of 1,400 new jobs which would be created by the new magnox facilities and THORP combined, about half were likely to be filled by local recruitment and half by immigration and that, of the half filled from local recruitment, about 230 were likely to be filled from unemployed and school leavers. I regard this estimate as realistic. If the 230 and the 700 are split between magnox and THORP on a proportionate basis the allocation to THORP would be 165 and 500. Spread over 10 years this would provide very little alleviation to the problem of unemployment. It would however provide some alleviation, and although it would be quite insufficient to justify the taking of any exceptional risk, it must be regarded as a point in favour of THORP. Mr Haworth for FOE WC accepted this when I put the matter to him. He said,

> 'I accept that it is certainly a plus, and this is to be welcomed in certain respects, but it is a very small plus, which is the point of our argument.'

Since I conclude that the effect upon employment will be beneficial, albeit not great, it follows that I reject the argument that THORP would have an adverse effect upon employment, or would aggravate the unemployment problem.

14.24 Before leaving the subject of employment I mention briefly certain other contentions:

i. that BNFL's demand for skilled labour would or might cause difficulties to other firms in the area. This contention I find to be without substance. Copeland made enquiries of the ten largest employers in the district. None considered the problem to be of any significance. Cumbria considered the history of BNFL's expansion as against closures of other firms in the area and were unable to find any link between the two.

ii. that BNFL was already a dominant employer. West Cumbria had in the past suffered from too great a dependency on a small number of dominant employers. It would be wrong to allow an expansion of BNFL's dominating position. This argument might have force if there were evidence that a refusal of permission would lead to a similar number of permanent jobs being provided by less dominant and equally stable employers. There was no evidence that it would.

iii. that the capital cost of each job created was inordinately high and that the money could be better spent on creating a larger number of jobs in spheres which did not carry any of the disadvantages of THORP. This argument too appears to be divorced from reality. No evidence was given that, if THORP were refused, money would be forthcoming to create the other jobs contemplated.

Training

14.25 BNFL have a long record of apprentice training schemes for school leavers. They have announced their intention:

i. to continue to co-operate with the local education authorities and the relevant Government agencies on training and education schemes.

ii. to pursue a recruitment policy aimed at local recruitment and keep the local authorities informed at regular intervals of the results.

iii. to assist financially with scholarship and other arrangements which in its judgment are sound and well suited to the needs of West Cumbria.

Traffic movements

14.26 Evidence on this subject, both as to the present situation and as to the likely results of the expansion, was presented in the form of a statement agreed between BNFL and the local authorities (CCC 35) and was not in dispute. The third part of the document covered recommendations for action. As to some of these, BNFL have announced their willingness to contribute towards, and take action on, certain road and other improvements not within the responsibility of the Department of Transport. They are as follows:

i. To finance:
   (a) an immediate improvement to the junction at Calderbridge on the A595 trunk road and the main access road to the Works by increasing the length of two-line approach to the trunk road to approximately 100 metres.
   (b) the improvement of the main access road except where it passes through Yottenfews to a standard allowing a minimum stopping sight distance of 90 metres.

ii. At such date as traffic flows in the main access road reach 1,000 passenger car units during the peak hours, to finance the construction of a local bypass at Yottenfews to a design standard of 60 km/hr.

iii. At such date as the Department of Transport carry out an improvement to provide a new at-grade junction between the trunk road A595 and the main access road to the Works, to finance the necessary new link from the trunk road to the main access road at Knocking Wood. (Estimated cost to BNFL at mid-1977 prices is approximately £136,000.)

iv. As soon as negotiations with land-owners have been completed, to finance the acquisition of land, the design, construction and supervision of construction of a new link road between the Works and the B5344 near Seascale which on completion to the satisfaction of the County Council will be taken over as a public highway. (Estimated cost at mid-1977 prices is £750,000 to £1 million.)

v. To implement a scheme to enable material presently transported by road from Windscale to Drigg to be conveyed by rail unless a feasibility study currently being carried out demonstrates such a scheme to be impractical or unreasonably expensive.

14.27 The remaining recommendations for action contained in the agreed statement CCC 35, which are all directly or indirectly the responsibility of the Department of Transport, are the following:

1. *Rail services*
   Every effort should be made to retain and improve the Cumbria coast line for both freight and passengers.
2. *Trunk roads A595 and A5092 from Greenodd to Workington*
   The Department of Transport should continue investigating what further improvements can be made, and whether or not improvements already programmed can be brought forward.
3. *The junction between the main access road (C4013) and the trunk road A595*
   The Department of Transport should be asked to consider improving the present trunk road so that a new necessary link from the trunk road to the existing main access road at Knocking Wood can be provided.

14.28 Regarding the first recommendation, the great majority of fuel flasks arriving at and despatched from Windscale travel by rail. I have no doubt that this should continue and that in so far as improvements are necessary to handle increased traffic they should be carried out. The flasks are of great size and weight and in the absence of a rail link would have to travel by road. In such circumstances extensive improvements might well be required to the road network both in Cumbria and near reactor sites. Railway passenger services however are presently used only by approximately 5 per cent of employees. Alterations to the service (which is not at present suitable for employees living to the north of the works) might increase usage a little, but it seems unlikely that the absence of the service would greatly increase the number of passenger vehicle movements to and from the works. I make no recommendation with respect to the maintenance or improvement of the passenger service. Both may well be desirable on other grounds but I am unable to see that the matter is significantly affected by THORP.

14.29 In connection with the second and third recommendations Copeland urged that a by-pass for Egremont and an improved link to the A66 should be constructed. I was assured by Mr Liddle that DTp were examining the possibility of small local improvements throughout the length of the A595, and in particular, whether or not THORP went ahead, to the Calder Bridge junction. He further assured me that in so far as THORP generated increased traffic this would raise such improvements higher in the Department's order of priorities. In the circumstances I see no reason to pursue this matter any further.

14.30 I am satisfied that road and rail traffic problems can be dealt with and constitute no ground for refusing permission.

Housing

14.31 The local authorities were satisfied that, even on the somewhat pessimistic assumptions (a) that half of the increased workforce would be immigrants and require accommodation and (b) that each immigrant would require a dwelling unit, ample land with planning permission existed (though requiring infrastructure works in some cases) to meet the anticipated demand. They were also satisfied that such demand would be within the capacity of the local construction industry. BNFL would require additional hostel accommodation. They have lodged an application for a new hostel to be built on the outskirts of Whitehaven and this is still under discussion with the local authority.

14.32 The local authorities are well seized of the importance of ensuring that housing for Windscale workers is sited in the most appropriate locations, and are anxious to avoid perpetuating earlier policies which led to the creation of company settlements. Copeland pointed out that, at present, not all the planning permissions for housing are in the locations which they regard as being the most appropriate. Subject, however, to the necessary infrastructure works, they were confident that demands could be satisfactorily met.

14.33 I heard no evidence which led me to conclude that I should reject the views of the local authorities that housing needs could be met. I therefore accept them.

Water supplies, sewerage and sewage treatment

14.34 It was accepted that the proposed development would create additional pressure on water supplies and sewerage both at the works itself and to satisfy the demand from the additional housing. As regards the provision of water to the works, the NWWA, BNFL and the County Council are satisfied that sufficient additional supplies could be made available although the sources remain still to be settled. I accept this and do not therefore examine this point further. As regards water supplies for additional housing the relevant authorities have concluded that, whether or not THORP is developed, some augmentation of supplies may well be required in the relevant areas, and that BNFL's expansion would have little effect on water resource development. Clearly then, water supplies are no obstacle to the proposed development.

14.35 The position regarding sewerage and sewage treatment is less satisfactory in that the NWWA acknowledge that existing standards leave a lot to be desired. So far as sewage from the works is concerned, BNFL have announced their intention to provide a new sewage treatment plant. So far as the treatment of domestic sewage is concerned, there are of course many competing claims for priority within the area of the NWWA. The Authority have, however, given an explicit undertaking that, if and in so far as it is necessary for health or safety to make any alterations in order to accommodate additional housing, this can and will be done. Moreover BNFL intend to meet the cost of an investigation into the condition and adequacy of sewerage and sewage disposal works in the areas of Copeland likely to be mainly affected directly or indirectly by the proposed development. I therefore conclude that, as in the case of water supplies, the position with regard to sewerage and sewage treatment gives no grounds for refusing planning permission.

Financing of housing and infrastructure improvements

14.36 I have referred in paras 14.25–26 and 14.35 above to the intentions which BNFL have announced to provide at their expense improvements in respect of roads and the transport of waste to Drigg, sewerage and training and employment. I should also mention that they are prepared, subject to the availability of suitable properties, to supplement the proposed new hostel and help meet the housing requirements associated with THORP by acquiring and using sites in the central area of Whitehaven to provide up to sixty housing units, as part of a comprehensive development scheme to be prepared by Copeland. It was put to me that it was unwise to rely merely on a statement of intent on the part of the company and that, where appropriate, binding agreements should have been made under Section 52 of the 1971 Act. I see no reason to doubt, however, that BNFL will abide by their stated intentions, and, since many of the infrastructure improvements are contingent upon the fulfilment of forecasts to which considerable uncertainty attaches, I do not consider that detailed and binding agreements would at this stage be appropriate.

14.37 It was argued by both local authorities that the exceptional nature of the proposed development made it inappropriate that additional expenditure on housing and infrastructure should have to be met from within existing programmes and ceilings at the expense of other projects: programmes should be expanded and central government assistance increased. As the Inquiry bears witness, THORP, if permitted, would in many ways be an exceptional development. However, the infrastructure expenditure for which additional assistance is sought would arise largely from the increased employment generated. Any important industrial development in West Cumbria would be likely to attract a proportion of the workforce from outside the area. In the normal course of events the local authorities would welcome such development and would be expected to meet their share of the resultant costs in the normal way. None of the evidence before me suggests that THORP would be in any way exceptional in respect of the numbers of jobs not filled by local people, and I therefore see no case for exceptional treatment as regards consequential expenditure on housing, infrastructure and related matters, other than those items which have already been agreed between the company and the local authorities.

Conclusions and recommendations on planning issues

14.38 My general conclusion is that, in planning terms, there is no substantial objection to the proposed development and that it would be likely to bring some positive benefits to the area. Indeed I am confident that had it been proposed to construct on the site a similar building, employing a similar number of employees and involving similar road movements, but making, for example, heavy machine tools instead of conducting processes involving the creation of plutonium and radioactive discharges, the proposal would have encountered no overall objection. The increased employment, the additional economic activity and the immigration into an area suffering the effects of emigration would all have been seen as outweighing any of the various matters I have considered under this section. I am satisfied that unless the more general matters raised in earlier sections afford – contrary to my views – a reason for refusal of permission there is no valid ground for refusal and that the case for permission has been fully established.

14.39 If permission were to be given, BNFL and the local uathorities accept that certain conditions should be attached. The following conditions are agreed between them:

1. (a) There shall be submitted to and approved by the local planning authority before the commencement of any particular part of the development to which this planning permission relates all details concerning the siting, design and external appearance of the buildings comprised in such part of the development, and the means of access thereto (hereinafter called 'the reserved matters').
   (b) Before submission of any such application for approval of a reserved matter, there shall be submitted to and approved by the local planning authority a layout plan showing the disposition of each of the buildings to which this permission relates, which disposition shall not differ substantially from that shown on the amended site plan No B24682D, and no development shall be carried out otherwise than in accordance with such approved plan.

2. The development to which this permission relates

must be begun not later than whichever is the later of the following dates:
   a. a date seven years from the date of this permission.
   b. the expiration of two years from the final approval of the reserved matters, or, in the case of approval on different dates, the final approval of the last such matters to be approved.
3. Application for approval of the reserved matters must be made not later than a date five years from the date of this permission.
4. A landscaping scheme shall be agreed with the local planning authority to be carried out on land in the applicant's ownership in the vicinity of the application site. The scheme shall incorporate a programme of implementation and a commitment to maintenance of any planting carried out in accordance with the scheme. Such scheme shall relate to that submitted and agreed in respect of permissions 4/76/0712 and 4/77/0200.
5. Any external plant above ground level shall be coloured in a manner to be agreed with the local planning authority.
6. The plant and buildings hereby approved shall not be brought into operation until necessary authorisations have been received from the Nuclear Installations Inspectorate and the Health and Safety Executive and the local planning authority shall be notified accordingly.
7. The applicant shall institute and maintain arrangements to be agreed with the local planning authority which will enable the local planning authority to be fully informed at agreed regular intervals about:
   i. the statutory controls which are applicable to the Windscale and Calder site;
   ii. the results of the environmental monitoring undertaken in pursuance of statutory requirements; including information on the composition and quantities of radioactive material discharged to the environment;
   iii. reports on all incidents affecting the Windscale and Calder site which are reportable to the Secretary of State for Energy.

14.40 Condition 1 does not take the usual standard form. Because the design and construction of THORP would be carried out in stages spread over 10 years, it has been so worded as to permit BNFL to seek detailed planning permission by stages rather than for the whole development at the same time. That apart, the conditions follow the lines of those attached to the permissions already given for the magnox part of the expansion plans and conditions 1, 2 and 3 are in line with standard statutory requirements. I recommend that the above eight conditions be attached in the event of permission being given for THORP.

14.41 A further condition was proposed by the County Council which would limit the size of THORP to a designed throughput of not more than 1,200 tonnes of spent fuel per annum. This would be acceptable to BNFL and I recommend that it should also be attached. It should be noted in this context that the application includes provision for HAWs whose capacity would be sufficient to accommodate the waste from only 3,000 tonnes of spent fuel. It is therefore possible that additional storage capacity may be required during the lifetime of THORP's operation but this will depend upon the rate of reprocessing and the speed of development of vitrification.

14.42 The County Council asked for the attachment of a further condition as follows:
'All irradiated fuel for reprocessing in the permitted buildings shall be delivered to the application site by rail, except in cases of emergency when rail facilities are not available.'
BNFL wished to amend this condition by the addition to the end of it of the words ' . . . and in the case of existing types of flask designed primarily for conveyance by road.'

14.43 Whilst I consider it desirable that as much transport of fuel flasks as possible should be by rail I do not consider that the proposed condition whether amended or not is an appropriate planning condition. I do not therefore recommend that it be included either in the proposed or amended form.

14.44 Finally it was suggested by the County Council that a condition should be attached requiring BNFL to provide facilities on such scale and in such manner as might be agreed for the measurement of radioactive body doses to the general public. I have made my recommendations as to the whole body monitoring in paragraphs 10.93–94 above. It is sufficient to say that here, as with other health and safety matters, I do not regard this as appropriate to planning conditions.

14.45 In conclusion, I therefore recommend that should permission for THORP be given it should be subject to conditions 1 to 7 set out in paragraph 14.39 above, and to the further condition as to size set out in paragraph 14.41 above.

15 The Inquiry Itself

15.1 Although it was not in the end suggested that there was any legal defect in the steps leading up to the Inquiry or in the Inquiry itself submissions were made that in a number of respects the Inquiry was unsatisfactory and that no decision on the application should therefore be made. Some of these I have already considered, e.g. that there should be an EIA prepared by BNFL or the local planning authority and an independent Inquiry into all past incidents at Windscale. I deal here with a number of other submissions made.

Time interval between the announcement of and the opening of the Inquiry

15.2 WA submitted that such interval was too short to enable objectors to prepare their cases. The application was called in on 25 March 1977 and the Inquiry opened on 14 June, some 11 weeks later, but the fact that the application would be called in was announced on 22 December 1976 and from that date it was known that there would be an inquiry. There were thus six months in which to prepare. I consider this period to be amply sufficient. WA pointed out that some objectors at any rate needed time to raise money before they could begin to prepare. Since some, including WA, were not formed until the call-in had been announced this is no doubt true but the proposition that time for preparation must be extended to enable groups of objectors to be formed and then to raise finance is in my view without foundation.

Location of the Inquiry

15.3 At the preliminary meeting it was submitted on behalf of WA that the Inquiry should not be held in the locality of Windscale but at some more central location. I then rejected that submission. Thereafter the matter of location was again mentioned from time to time. I appreciate fully that the nature of the application was such that it was of great concern not merely to those living in the area but also in the country at large and indeed in other countries. The attendance of witnesses from America and Japan and of the press and television from other countries speaks for itself. I have, however, no doubt that, even if the law were to permit such an Inquiry to be held in a place far removed from the locality to which the application relates, it is desirable that it should be held locally. A large number of witnesses and individual objectors were local residents. In addition a considerable number of local residents attended the Inquiry from time to time. Local residents are the people most immediately affected by the result of the application and should be able to attend and observe what is occurring. Much of the suspicion and alarm which exists stems from lack of information and a feeling that decisions are being taken by people in far away places on the basis of information, or the lack of it, unknown to those most immediately affected. Had the Inquiry been held in a more central location it would of course have saved some objectors expense and inconvenience but it would have caused extra expense and inconvenience to those most immediately concerned and affected, would have deprived some of them of the opportunity to attend and would probably have deprived me of some of what I regard as being the most valuable evidence tendered. This aspect of the matter must, I think, have escaped the attention of those who found cause for complaint in the location of the Inquiry. There is a curious inconsistency between advocating greater public participation and at the same time seeking a location which would reduce, or make more difficult and expensive, the participation of those principally affected.

Programming

15.4 Complaints of a somewhat querulous nature were made about the inconvenience caused by programme changes and about general sitting hours. I mention them for completeness only. As to sitting hours they were from 2 pm–5 pm on Mondays, 10 am–1 pm and 2 pm–5 pm on Tuesdays to Thursdays, and from 9 am–11.45 am on Fridays. These were long hours and imposed a considerable workload on all concerned. There was no substantial adjournment. I see however no reason to suppose that they rendered any objector unable to present his case properly. This particular matter was raised by WA. It is true that from time to time they were not fully up to date with what had transpired but this appeared to be due not to the pressure of work but to the taking of holidays.

15.5 As to programming, inconvenience is from time to time inevitable, since it is impossible to forecast precisely how long a particular witness or witnesses will be in the witness box. Those in charge of programming made great efforts to meet the convenience of everyone, as did the various parties between themselves, and the general principle adopted was to keep the programme as flexible as possible. As the Inquiry neared its end there was necessarily less room to be flexible. This matter was principally raised by Mr John Tyme of the Society for Environmental Improvement (SEI). This was somewhat curious for, entirely to meet their convenience, a group of SEI witnesses were, on the application of Mr Tyme interposed to give evidence during the course of BNFL's case. Since most parties in their closing speeches paid tribute to the Inquiry Secretary and his staff for their co-operation and assistance I cannot regard this complaint as being of substance.

15.6 I should, however, mention the one major change in programme which occurred. At an early stage I decided to complete all evidence on the wider issues before taking evidence on what may be termed conventional planning issues. The taking of evidence on the wider issues, however, occupied so much time and involved the absorption of so much new technical material daily that I in fact heard the planning evidence before completing the evidence on the wider issues. This was in part to enable me to keep abreast of new material and in part to enable my assessors to attend to urgent outstanding matters in their normal spheres of occupation. One technical witness from overseas, whose arrangements could not be changed, was in fact heard during this period. This change of programme was discussed with the parties, including the objector calling the technical witness, and was agreed before it was put into effect. My assessors read the transcript of the technical evidence thereafter so as to be able to advise me upon it.

The identity of my assessors

15.7 Mr Tyme submitted that this report would carry less weight because my assessors were, in the view of his clients, associated with the nuclear industry. I record this for information and because it reveals the state of suspicion which exists in certain people. Since Sir Edward Pochin is known for his work over many years in connection with radiological protection and Sir Frederick Warner for his work in connection with anti-pollution I found it difficult to understand the submission.

Financial disparity

15.8 Both before and during the Inquiry great stress was laid by a number of objectors on the disparity between the resources available to the applicants and to the objectors. That this disparity exists is clear. Despite its existence, however, the oral evidence tendered and the documentary evidence submitted was very wide-ranging and highly qualified. Supplemented as it was by work done and evidence given at my request I am satisfied that it was fully adequate for the purpose of making a decision on the application.

15.9 There can nevertheless be no doubt that the costs of presenting a fully developed case at the Inquiry and, equally important, investigating the validity of the applicant's case, are very considerable. I say *a* fully developed case because although no objector covered every matter they did between them cover, so far as I was able to ascertain, all matters. That a fully developed case should be presented is plainly in the public interest and it is possible that the drain on resources caused by the presentation of such a case at this Inquiry could prejudice the presentation of such a case, and thus the public interest, at any Inquiry in relation to CFR I. This point was specifically mentioned on behalf of FOE and WA who doubted their ability to mount a case at another similar Inquiry. I can make no recommendation on the matter. I draw attention to it for it appears to me to be one which should be considered.

Costs of the inquiry

15.10 I was not asked to make any recommendation with regard to orders for costs under Section 250(5) of the Local Government Act 1972 nor, had I been asked, would I have seen fit to do so. No party was guilty of any conduct which would justify a recommendation for an adverse order. No party or parties could be said to be deserving of some particularly favourable treatment.

15.11 Cumbria did, however, ask that I should report, and I do so, a submission that the costs of providing the Inquiry accommodation at the civic hall, the telephones and the sound equipment, which they had been required by DOE to bear, ought to be borne nationally rather than locally, since the Inquiry had been concerned largely with matters of national policy and interest.

Nature of the inquiry

15.12 It was submitted on behalf of WA that the Inquiry should have been of an investigatory rather than an adversary nature. This submission I report. I add that, although the Inquiry procedure is in general of an adversary nature I regarded it as my duty to, and accordingly did, take steps to investigate any matter which appeared to me or to either of my assessors, to require investigation whether or not it had been raised by any of the parties.

15.13 It was further submitted that it would have been more satisfactory to hold a Planning Inquiry Commission rather than a public local inquiry. Had specific alternative sites been seriously suggested, and had there been no existing facilities at Windscale and no store of information covering discharges therefrom there might have been considerable force in this submission but in the prevailing circumstances it appeared to me of little merit.

15.14 A final matter deserving of mention is that much time was occupied after the Inquiry had opened in objectors seeking further information from BNFL and in the provision of such information by BNFL. With such complex subject this is to some extent inevitable.

An exploratory question in cross-examination frequently opens up a whole new field of investigation. It is possible that had there been more preliminary investigation some time might have been saved but it is by no means clear that this is so. Before the Inquiry opened BNFL had made available to objectors a vast amount of material. Their written case and lists of documents were, pursuant to a direction under Rule 6(6), served on 11 May, 6 weeks before the Inquiry, the documents themselves were made available for inspection in Whitehaven and London respectively on 21 and 25 May 4 and 3 weeks before the Inquiry, proofs of evidence of 8 of their principal witnesses were served 2 weeks before the Inquiry opened, and the remainder on the first day of the Inquiry. Cumbria's statement of case and list of documents were served on 13 May in accordance with Rules 6(2) and 6(4). I am satisfied that the information provided to objectors before the Inquiry was, save as to the financial aspects upon which I have already commented, as full as anyone could reasonably expect. In the course of the Inquiry BNFL further provided with great willingness almost all information for which they were asked and to refer, as did one witness, to information being extracted from them 'drop by reluctant drop' I can only describe as absurd. In certain cases they were, it is true, reluctant. In such cases their reluctance was for the most part entirely reasonable. In those few cases where it was, in my view, not reasonable I required them to supply either the information sought or such of it as appeared to me to be sufficient to serve the objector's purpose.

15.15 I am satisfied that there is no reasonable cause of complaint by objectors and this was accepted by most of them. Indeed more than one paid tribute to the assistance they had received from BNFL.

16 Overall Conclusion and Recommendation

16.1 My overall conclusion is that outline permission should be granted subject to the conditions set out in paragraphs 14.39 and 14.41 above and that such permission should be granted without delay.

16.2 I am authorised by my assessors to say that they agree both with my overall conclusion expressed above and with all subsidiary conclusions save those relating to conventional planning issues for the hearing of the evidence upon which they did not attend, and save also those relating to questions of law.

17 Summary of Principal Conclusions and Recommendations

Conclusions

17.1 It is convenient to summarise my conclusions by way of giving my answers to the three questions set out in paragraph 1.7 and the principal reasons which have led me to arrive at such answers. This will necessarily involve some repetition of what has appeared before but this is unavoidable.

Question 1. Should oxide fuel from UK reactors be reprocessed in this country at all?

17.2 Although reprocessing of oxide fuel is not necessary to preserve the option either to build CFR1 or to launch an FBR programme, and although it is possible that it will be decided not to proceed further with FBRs at any rate for a period, I conclude that a new plant for reprocessing oxide spent fuel from UK reactors is desirable and that a start upon such a project should be made without delay. My principal reasons for this conclusion are as follows:—

1. Stocks of spent fuel from AGRs presently existing and under construction will, unless reprocessed, continue to build up and will have to be stored until finally disposed of in some manner.
2. It is necessary to keep the nuclear industry alive and able to expand should expansion be required. Such expansion might be required, either to meet additional energy demands, or to preserve a 'mix' and to avoid over-dependence on a particular energy source, or to reduce the number of fossil fuelled stations as a result of confirmation from further research of the views expressed in the Ford Foundation Report (and elsewhere) that such stations are more harmful than nuclear stations.
3. Keeping the industry alive will involve further reactors being constructed and further quantities of spent fuel arising. Such further quantities will, if not reprocessed, also have to be stored until finally disposed of in some manner.
4. All the spent fuel stored will contain fission products and the long-lived actinides including plutonium. The inventory of plutonium will therefore continue to increase for so long as reprocessing is delayed.
5. The prolonged storage of ever-increasing spent fuel containing an ever-increasing quantity of plutonium would involve the development of new storage methods. This would be both a costly and a lengthy process.
6. To store such increasing quantities of spent fuel would only be sensible if it was likely that it would ultimately be decided to dispose of the spent fuel (with its entire content of plutonium and other radioactive substances) without reprocessing.
7. Such a decision appears to be unlikely and not to be in the best interests of ourselves or future generations. This is because:
    i. It involves throwing away large indigenous energy resources and, for so long as there is a nuclear programme of any kind, making us wholly dependent on foreign supplies. The undesirable consequence of energy dependence of this nature has been only too well demonstrated in recent years in the case of oil.
    ii. It involves committing future generations to the risk of the escape of more plutonium than is necessary. If the plutonium is extracted by reprocessing the total inventory can be greatly reduced.
    iii. It involves committing future generations to a greater risk of escape of the remaining content of the spent fuel since the spent fuel is likely to be more vulnerable to leaching by water than solidified highly active waste.
8. If reprocessing is going to take place at some time it is preferable to start without delay since the techniques can then be developed at a reasonable rate, and greater experience can be gained, both of the process itself and of the behaviour and effects of the emissions involved, whilst spent fuel stocks and arisings are comparatively small. This is to the benefit of workers, public and future generations alike.
9. The risks from the emissions involved in reprocessing are, on current estimates, likely to be very small and, if reprocessing is to take place at some time, will in any event occur at some time. Evidence that current estimates are seriously wrong did not appear to me to be convincing but, should it be proved correct, this is likely to have occurred well before THORP begins to operate. THORP would then have to operate to the new limits or not at all.
10. The risks of accident will, if reprocessing is to take place at some time, also have to be incurred, at some time. At the present time they are likely to be containable within tolerable levels. If reprocessing were to begin suddenly on a large scale after a lapse of time the risks would probably also be containable but would be likely to be greater.

11. The risks from terrorism are not significant. The plutonium separated from UK fuel would be stored at Windscale and would not be subjected to movement from Windscale save in the form of fuel, which is not an attractive target.
12. The risks arising from transport would be no greater than at present. Spent fuel will have to be carried to Windscale in any event. Fresh fuel sent out from Windscale would not present any significant risk.

Question 2. Should reprocessing be at Windscale?
17.3 I have no doubt that the answer to this question should be in the affirmative. The existence of the facilities already at Windscale and the store of knowledge concerning the behaviour of radionuclides discharged from Windscale, coupled with the facts that any alternative would be likely to involve additional transport of plutonium or prohibitive expense, make it clear that, if the operation is to be carried on at all, Windscale is the obvious location. It will involve additional exposure to local inhabitants but the risks involved appear to me to be so small that this fact cannot outweigh the advantages mentioned.

Question 3. Should the plant be double the size required for UK spent fuel and used to reprocess foreign fuel?
17.4 The financial advantages of having a plant to reprocess foreign fuel on the basis intended by BNFL are plain. There is the additional advantage that planning permission, a start on THORP and the receipt of foreign fuel for reprocessing would do something to relieve the pressure on non-nuclear-weapon states to develop their own facilities. It would also demonstrate that this country intends to honour at least the spirit, and as I think the letter, of its obligations under the NPT. This could well be an advantage in negotiations, over the period when THORP is building, to strengthen the NPT. Furthermore, the existence of substantial reprocessing facilities in one or more nuclear-weapon states is a necessity to deal with fuel which fails in reactors or deteriorates in storage.

17.5 The disadvantages of accepting and reprocessing foreign fuel are also clear. It will involve additional routine emissions, additional storage of spent fuel pending reprocessing, additional highly active waste to dispose of and, which was chiefly relied on, additional movements of plutonium in some form, and the putting of non-nuclear-weapon states nearer to the bomb.

17.6 These disadvantages appear to me to be clearly outweighed by the advantages. The risks from the additional routine emissions are very small; the additional storage presents no significant risk and certainly no greater risk than would be involved in the storage for prolonged periods of UK spent fuel; the total highly active waste from reprocessing of UK and foreign fuel combined will contain only a fraction of the plutonium which would be contained in UK fuel alone if such fuel were disposed of without reprocessing; the risks from the movement of plutonium can be largely dealt with by technical fixes. The one substantial objection which appeared to me to arise is that the separation of plutonium and its supply to non-nuclear-weapon states will put them nearer to the bomb. Since, however, this matter can be alleviated to some extent by technical fixes; since it will not in any event happen for 10 years; and since a refusal to accept foreign fuel would be in breach of the spirit if not the letter of the NPT and would put pressure on non-nuclear-weapon states which could lead them to produce their own plutonium long before they could receive any from THORP I cannot regard this as an overriding objection.

17.7 It is also important to remember that unless foreign business on the required scale can be obtained BNFL would not proceed with the plant as presently proposed. To meet UK needs only would require a smaller plant and the whole concept would have to be the subject of reconsideration and re-design. This would be likely to involve an undesirable delay in starting on reprocessing of UK fuel. It would also mean that when further capacity was required we should, instead of having it available at the cost of foreign customers, have to finance it ourselves.

In the light of the above I would answer the third question in the affirmative.

Recommendations

17.8 My principal recommendations are the following:
1. Consideration should be given to charging some independent person or body with the task of (a) vetting security precautions both at Windscale and during transit of plutonium from Windscale and (b) reviewing the adequacy of such precautions from time to time (para 7.18).
2. BNFL should devote effort to the development of plant for the safe removal and retention of krypton 85 and, if development proves successful, should incorporate it in the proposed plant (para 10.52).
3. More permanent arrangements for whole body monitoring of local people should be instituted. Subject to certain general principles, the details should be agreed by those directly concerned. They would not be appropriate to planning conditions (paras 10.93, 10.94 and 10.126).
4. The authorising departments should however consider whether provision of such facilities should be made a condition of authorisations to discharge (para 10.95).
5. Consideration should be given to the inclusion of some wholly independent person or body with environmental interests in the system for advising central government on the fixing of radiological protection standards. That person or body should probably be changed from time to time (para 10.111).
6. A single Inspectorate, as recommended by the Royal Commission, should be responsible for

determining and controlling all radioactive discharges (para 10.113).

7. There should be specific discharge limits for each significant radionuclide. The onus should be placed clearly on the operator to show that a discharge cannot practically be avoided before the limits are fixed (paras 10.115–10.116).
8. The provisions of the Radioactive Substances Act 1960 relating to the powers to hold inquiries into proposed authorisations to discharge should be re-examined (para 10.122).
9. The relevant authorities should carry out more monitoring of atmospheric discharges (para 10.126).
10. FRL should publish their annual reports more rapidly in future. There should, as recommended by the Royal Commission, be one comprehensive annual survey published of all discharges and at intervals, reports by NRPB on radiation exposure (para 10.126).
11. BNFL should do more, in future, to ensure that safety precautions and operating procedures at Windscale are sufficient for all eventualities, are strictly observed and are continually rehearsed. (para 11.11.)
12. The current review of NII should examine whether they are sufficiently equipped with scientific expertise to check the designs for the proposed plant (para 11.24).
13. It is essential that those who would be required to take action under the Windscale emergency plan are fully aware of the responsibilities the plan places on them (para 11.30).
14. The local liaison committee should be re-organised and its functions re-defined. (para 11.34).
15. Fuel flasks should, as far as possible, continue to be delivered to Winsdcale by rail, but this is not a matter appropriate to planning conditions (paras 14.28 and 14.45).
16. Outline planning permission for THORP should be granted without delay, subject to conditions (paras 14.39–14.41, 14.45 and 16.1).

18 Miscellaneous Matters

Inquiry procedure

18.1 Owing to its special nature I made certain departures from ordinary procedure. I list them here with the reasons for their adoption:

i. *Brief opening speeches by all parties before the applicants fully opened their case*

This was partly to enable me and my assessors to obtain as early as possible a general picture of the issues likely to arise. For this purpose it was invaluable.

It was also to ensure that objector's points should be made public at the outset instead of having to wait for many weeks for expression. As far as I know this purpose was achieved.

ii. *All evidence on oath*

This appeared to me desirable in view of the conflict of evidence which appeared likely to arise and the suspicion which appeared to be felt by objectors of statements made by BNFL. Happily, in the event, the suspicion was to a great extent dissipated in the course of the Inquiry.

Written representations

18.2 In all, 161 such objections were received and all have been taken into account. Where they raised technical medical or scientific matters they were also seen by and as necessary discussed with my assessors.

Acknowledgements

18.3 I have already expressed my thanks to some of those who contributed greatly to the Inquiry. In addition I wish to express my gratitiude to the staff of the Civic Hall at Whitehaven upon whom the Inquiry imposed a great burden, to the sound engineers, who never failed to ensure that everyone could hear, to Mr John Rhodes, the Inquiry Secretary, and his staff, whose organisation and assistance to myself, to my assessors and to the parties was of the greatest value and lastly to my assessors themselves to whom I owe a great debt for their unfailing and patient assistance both during the period of the Inquiry and in writing this report, and whose wide knowledge enabled me to raise and explore during the Inquiry some points which had escaped even the fine mesh of the objectors' net.

Annex 1 –

List of Appearances

Annex 2 –

List of Documents

are contained in a separate volume

Annex 3

Somatic Risk Estimates

Estimates are given in para 10.35 of the frequency with which fatal malignant disease might be induced by radiation from operation of the THORP plant, with or without release of krypton 85. These estimates are based on:
a. the collective doses attributable to THORP as derived from documents G57 and BNFL301 for atmospheric, and BNFL241 for aqueous discharges, from which annual dose commitments (man-rem per year of discharge) can be derived.
b. the multiplication of the man-rem value for each radionuclide according to the weighting factor, as given in ICRP Publication 26, appropriate for the body tissues irradiated by that isotope, to give the relative effectiveness of each collective dose in inducing fatal malignant disease.
c. the addition of collective doses from occupational exposure, as given in BNFL302 for the THORP plant.
d. the total from all these sources, as shown in the Table below, equivalent to a whole-body collective dose commitment of 21,900 man-rem per year of operation. This figure when multiplied by the estimate given by ICRP (publication 26) of 10^{-4} fatal cancers per man-rem of whole-body irradiation, corresponds to a rate of 2.2 such cancers per year of operation, if cancers are induced at this rate by these low incremental doses. If krypton 85 were not being discharged this figure would be reduced to 0.9.

Collective Doses Due to Discharges (with margins) and Occupational Exposures from THORP

Nuclide	Document reference	Collective dose per year of reprocessing (man-rem)	Tissues or organ irradiated	Equivalent whole-body dose (man-rem)
Aqueous discharges				
Cs134/137	BNFL241	2,141	whole-body	2,141
Sr90	BNFL300	1,600	bone	64
Pu (soluble)	,,	1	bone	0
Atmospheric discharges				
C14	G57	5,400	whole-body	5,400
H3 (tritium)	G57	450**	whole-body	450
I129	G57	2,800**	thyroid	112
Sr90	BNFL301	300	bone	12
Pu (if soluble)	BNFL301	2,650	bone	106
[if insoluble	BNFL301	76	lung	20]
Kr85	G57	1,750	whole-body	1,750
Kr85	G57	370,000	skin*	11,100
Occupational	BNFL302	800	whole-body	800
			Total	21,935

Collective dose commitments are given per year of reprocessing (assuming 10 years of operation, with integration of dose commitments over 100 years) as quoted in the reference documents cited. Equivalent whole-body doses are derived in respect of fatal cancer induction according to ICRP weighting factors for individual tissues (its publication 26, para 105), relative to a fatal cancer risk of 10^{-4} per rem for uniform whole-body radiation (pubn. 26, para 60) corresponding to the non-genetic component of the whole-body weighting factor.

*For skin, ICRP consider that the risk of fatal radiation induced cancer is 'much less' than for other tissues considered (pubn. 26, para 63) including bone and thyroid. Here however a weighting factor of 0.03 relative to fatal cancer from whole-body radiation is applied (e.g. as compared with 0.03/0.75 = 0.04 for bone and thyroid).

**Includes dose from aqueous discharge.

Sir Edward Pochin

Annex 4

Genetic Risk Estimates

The estimated frequency of substantial genetic abnormalities, given in para 10.36 is based on the following estimates:

a. Whole-body collective dose commitments (per year of reprocessing by THORP, with dose commitments integrated to 100 years) are given in documents BNFL241 and G57 for radionuclides causing gonadal irradiation, and G57 gives the relationship between gonadal and mean whole-body dose for different nuclides.
b. Occupational exposures from THORP are estimated in BNFL302.
c. Estimates of $3 \cdot 10^{-4}$ substantial genetic abnormalities per genetically significant man-rem are given in the 1972 BEIR and the 1972 UNSCEAR reports, and of $2 \cdot 10^{-4}$ in the 1977 UNSCEAR and ICRP reports.
d. Taking account of the age structure and mean ages of conception, about 40 per cent of the collective dose to the general public, and 30 per cent of that to a population of occupational ages, is estimated to be genetically significant.
e. The following table indicates a total of about 3,500 man-rem of genetically significant exposure per year of practice, corresponding to the causation – in the total of all subsequent generations – of 0.7 to 1.0 substantial defects per year of operation, on the basis of the two risk estimates given in para c above.

Collective doses (man-rem) per year of operation of THORP

	Whole-body		Gonad / Whole-body		Gonad		Genetically significant fraction		Genetically significant dose
Aqueous									
Cs 134/137	2,141	×	1.0	=	2,141	×	0.4	=	856
Atmospheric									
H3	450	×	1.0	=	450*				
C14	5,400	×	0.6	=	3,240				
Kr85	1,750	×	1.2	=	2,100				
					5,790	×	0.4	=	2,316
Occupational	800	×	1.0	=	800	×	0.3	=	240
								Total	3,412

*includes dose from aqueous discharge.

Sir Edward Pochin

Annex 5

Mean Genetic Dose to Local Population

The genetically significant irradiation of populations living in the proximity of Windscale, from the proposed waste discharges, can be estimated on the basis of:

a. The estimated collective doses to populations within 50 km of Windscale from all radionuclides involving significant whole-body or gonadal irradiation. Values are given for collective dose commitments per year of reprocessing (as integrated for 100 years), in G57 for atmospheric, and in BNFL241 for aqueous, discharges that may result from a refurbished magnox and from THORP.

b. The relationships between gonad dose and mean whole-body dose for relevant isotopes as given in G57.

c. These values indicate a total (gonad) collective dose rate of 230 man-rem per year within the population as follows:

Population within 50 km
Annual collective whole-body dose

Radionuclide	Refurbished magnox	+	THORP	=	Total		Gonad dose / Whole-body dose		Annual collective gonad dose
Krypton 85	0.5	+	4.4	=	4.9 man-rem	×	1.2	=	5.9 man-rem
Tritium	2.6	+	8.6	=	11.2 ,,	×	1.0	=	11.2 ,,
Carbon 14	5.7	+	9.8	=	15.5 ,,	×	0.6	=	9.3 ,,
Caesium 134/137	145	+	58	=	203 ,,	×	1.0	=	203 ,,
									229 ,,

d. The population within 50 km is estimated as 300,000 (BNFL230). An annual collective dose of about 230 man-rem distributed within this population implies an average gonad dose per person of 23 millirem per 30 years, or one-quarter of the limit of 0.1 rem per 30 years indicated in Cmnd 884 as applying to the whole population from waste disposal.

Sir Edward Pochin

CONSTRUCTION LIBRARY,
Liverpool Polytechnic,
Victoria Street, L1 6EY